New Directions
in Sex Research

PERSPECTIVES IN SEXUALITY
Behavior, Research, and Therapy

Series Editor: RICHARD GREEN
State University of New York at Stony Brook

NEW DIRECTIONS IN SEX RESEARCH
Edited by Eli A. Rubinstein, Richard Green, and Edward Brecher ● 1976

A Continuation Order Plan is available for this series. A continuation order will bring delivery of each new volume immediately upon publication. Volumes are billed only upon actual shipment. For further information please contact the publisher.

New Directions in Sex Research

Edited by

ELI A. RUBINSTEIN and RICHARD GREEN

State University of New York at Stony Brook

and

EDWARD BRECHER

PLENUM PRESS • NEW YORK AND LONDON

Library of Congress Cataloging in Publication Data

Main entry under

New directions in sex research

(Perspectives in sexuality)
Proceedings of a conference held at the State University of New York at Stony
Brook, New York, June 5-9, 1974 and jointly sponsored by the Dept. of Psychiatry
and Behavioral Science at SUNY, Stony Brook, the Institute for Sex Research at
Indiana University and R. Green.
(The material in this book originally appeared in Archives of sexual behavior, vol. 4,
July, 1975.)
Includes index.
1. Sex research–Congresses. I. Rubinstein, Eli Abraham. II. Green, Richard, 1936-
III. Brecher, Edward M. IV. New York (State). State University at Stony
Brook. Dept. of Psychiatry and Behavioral Science. V. Indiana University. Institute
for Sex Research. VI. Archives of sexual behavior.
H.Q.60.N48 1976   301.41'7'072   76-14490
ISBN 0-306-30956-4

Proceedings of the Conference held at the State University
of New York at Stony Brook, New York, June 5-9, 1974

The material in this book originally appeared in
Archives of Sexual Behavior Vol. 4, No. 4, July, 1975.
First published in the present form by Plenum Publishing Corporation in 1976.

© 1975 Plenum Publishing Corporation
227 West 17th Street, New York, N.Y. 10011

Printed in the United States of America

Preface

This publication comprises the proceedings of a conference held at the State University of New York at Stony Brook, Stony Brook, New York, June 5 - 9, 1974. The conference was jointly sponsored by the Department of Psychiatry and Behavioral Science at SUNY, Stony Brook, the Institute for Sex Research at Indiana University, and the Editor of this journal. Financing for the conference came from the National Institute of Mental Health.

The initial planning for the conference was a shared effort of Stanley F. Yolles, Paul Gebhard, Richard Green, and Eli A. Rubinstein. In addition to the planning of the conference and the selection of participants, all four served as program coordinators during the conference to help ensure productive use of the limited time available. Grateful acknowledgment is given for the advice and consultation of Jack Wiener, George Renaud, and Betty Pickett of the NIMH, who provided assistance in the development of the conference plans.

The selection of participants was planned to permit a wide sampling of researchers working in areas of significance to the future of sex research. The titles of the papers listed in the table of contents reveal this broad approach. Furthermore, the remaining participants were included as individuals whose research and scholarly interests would ensure a multidisciplinary approach to the topics under discussion.

The individual presentations, in addition to their substantive contributions, provided stimulation for the group discussions, which are summarized following each formal paper. A combined reference list for all the formal papers is included at the end of this publication.

Editing was a challenge we hope was successfully met. Edward Brecher was invited to the conference as a professional writer who has distinguished himself for his highly competent publications in the area of sex research. It was his responsibility to put the proceedings into its first semblance of order. Thereafter, he and the other two editors worked closely in putting the document into its final form.

At the time the conference was in development, a number of individuals at Stony Brook and elsewhere expressed interest in attending the meeting. To ensure maximum interaction among the invited participants, the conference was limited to the conferees. We hope that this publication will serve as a useful means of now extending the conference to the larger audience. More importantly, we hope this publication can serve as a marker for the future direction of sex research.

E. A. R.

R. G.

Contents

Opening Address[1]

Stanley F. Yolles, M.D. [2]

MY FRIENDS AND COLLEAGUES:

As Chairman of the Department of Psychiatry and Behavioral Science, I am delighted to welcome you to Stony Brook — especially so because we are here as participants in what may well be a significant "first."

It appears that sex research — like some kinds of sexual behavior — has come out of the closet. Therefore, as leaders in research in human sexuality, we have a responsibility not only to discuss what the future directions of research in the field *should* be but also to take that next important step in developing a comprehensive and coordinated program.

Coordination itself should be comprehensive. For example, this department is named Psychiatry and Behavioral Science to identify the scope of our program. It is increasingly evident — certainly in the academic world — that medicine and the behavioral sciences are inevitably linked, as we attempt to apply the knowledge we already have in bettering the human condition.

This linkage is particularly significant in consideration of human sexuality. Our interests and responsibilities now go beyond research to include teaching, treatment, and the establishment of a sound policy on which to base our work, and, most important, to decide what that work will consist of, while setting priorities for the immediate future.

The only way we can hope to achieve these objectives is through collaboration, a commodity in short supply in this field in the past. It was in recognition of this need for joint planning and a consensus that this workshop was organized. It, in itself, represents the beginnings of a collaboration that can become national in scope.

---

[1] Presented at the conference, "Sex Research: Future Directions," held at the State University of New York at Stony Brook, Stony Brook, New York, June 5-9.
[2] Professor and Chairman, Department of Psychiatry and Behavioral Science, State University of New York at Stony Brook, Stony Brook, New York 11794.

As you know, arrangements for this meeting were coordinated by the State University of New York here at Stony Brook, the Institute for Sex Research at Indiana University, and the Editor of the *Archives of Sexual Behavior,* and with the support of the National Institute of Mental Health.

Our discussions here will be concentrated on research; but, if we do our work well, what we decide here will inevitably influence the world outside the laboratory, as well as chart the course of future research investigations.

Given this objective, the workshop coordinators have planned the sessions in the expectation that this meeting will *not* be concentrated on another recital of the state of the art of research in human sexuality. The literature is extensive, if one wishes to read what has already been done in the field. Our task is to discuss what we do next, and how we improve research procedures.

Participants will each have the dual role of teacher and student, so it is a matter of enlightened self-interest to go beyond the usual routines of an exchange of information and to work as a committee of the whole, in considering process and substance, and *then* to make recommendations which will be considered by the National Institute of Mental Health in planning support priorities in this field.

What we will be doing here, in essence, will be to help in establishing a public policy in sex research. This is the first time any group outside the government has had such an opportunity, and it will not be easy.

Research in human sexuality proceeds through a maze of beliefs, attitudes, values, actions — and *some* facts. More than in any other field, human sexuality brings to the surface major value conflicts existing in our society, with no consensus among adults. The intensity of these conflicts makes it exceedingly difficult for research investigators to select and analyze data in what Bob Stoller (1973) terms "the slow, steady, patient, careful exploration of reality that typifies the scientific method."

In today's range of sexual attitudes and behaviors, it is even difficult to define "reality." But we are here to give it a try.

It might be helpful, for the record, to remind ourselves that there have been (and still are) other areas surrounded by taboos, prohibitions, controls, and professional fears, and that professionals in those fields, too, have been forced by events and pressures to establish policies and procedures. We might well keep this in mind and, perhaps, learn something from their mistakes to apply to our thinking and discussion here. In 1958, for example, research in psychotherapy had hardly begun to explore the full complexity of the therapeutic process. So, in an effort somewhat comparable to this workshop, the National Institute of Mental Health sponsored the first national conference on Research in Psychotherapy, held in Washington, D.C.

The proceedings, edited by Rubinstein and Parloff (1959), contain this comment, as an illustration of the timid and self-protective stance of at least some psychotherapists at that time: "The responsiveness of some patients to psychotherapy is suspected to involve a strong element of suggestion. To confirm this would be to imply that the psychotherapist does not have a *unique* contribution to make in the treatment of patients. It was, therefore, too dangerous to study."

Well, the perception of danger on the part of those therapists was, as it turned out, overweening. I think it can be demonstrated that in the years since 1958, psychotherapists by and large have come to agree and admit that they do *not* have a *unique* contribution to make in the treatment of patients but that they have a significant contribution to make. Their numbers and prestige continue to increase, while their practice benefits more people.

In all honesty, however, it must be admitted that other indications of protective attitudes continue to be manifest. Studies of the mind — like studies of human sexuality — are disquieting to the public. But they don't burn witches today in America, and in the current climate of sexual mores it appears that perception of danger among sex research investigators need not be acute. This is not to say that researchers have not been subjected to some very clear and present dangers at various times — and not too long ago.

In research on narcotics and other illicit drugs, for example, the investigator (for years) was confronted with a combination of dangers. They ranged from benign neglect to personal harassment. Neglect resulted in far less than adequate funding for drug research. And the neglect was benign because, generally speaking, the public concerned itself only with narcotic addicts who committed crimes; and even that concern was not widespread. Furthermore, there was little stimulus from within the medical profession for drug research. The major medical point of view toward narcotic addiction was characterized by almost complete uninterest in any segment of addiction itself or the problems resulting from addiction.

There were, luckily, some notable exceptions, so that when public, professional, and political panic about drug abuse brought about increased funding for drug research a body of knowledge *did* exist, on which drug investigators could build.

But it must not be forgotten that drug research has entered the political arena and is therefore subject to uneven pressures and a demand for instant solutions to age-old problems. This should serve as a warning to those who seek to establish a comprehensive design for research in human sexuality. The field is already tainted by faddism, and is subject to pressures that I would like to discuss in a moment.

But first let us consider the danger of harassment of investigators. This has been fairly constant in the drug field and to an extent still exists. Law enforcement officers and others within the system of criminal justice sometimes made the life of research investigators so unpleasant — regulations, sudden searches of laboratories, difficulties in securing the drugs themselves — that many gifted scientists simply refused to enter the field.

This situation, however, has improved greatly, and this fact, coupled with the developing technology in drug research laboratories, has given drug research a respectable status and provides absorbing challenges for some of the best minds in the fields of pharmacology, the basic sciences generally, and increasingly the behavioral sciences.

The situation relating to research in alcohol abuse and alcoholism has been characterized by some of the same difficulties. Additionally, however, the situation *vis-à-vis* alcohol research developed some self-perpetuating problems of its own. No one will argue the point, I feel sure, that alcoholics are among the most difficult patients to treat. As individual persons, they are among the most frustrating to physicians and other clinicians. So again, until very recently, most practicing physicians felt that concern with alcoholism was best left to a few crusaders and that it, like narcotic addiction, was not a part of a respectable and profitable practice of scientific medicine.

These attitudes were transmitted to the research community, so that — again with some major and brilliant exceptions — research in alcohol did not attract the best and most highly qualified investigators. This, too, is changing and the quality of alcohol research now reflects the changes.

These conditions, while not identical to conditions hampering valid research in human sexuality, are analogous in large extent. One difficulty in sex research, as far as public attitudes are concerned, is the knowledge that people in modern society — from the Victorian period to the very recent past — have projected and supported a public view of sex that has traditionally had little relationship to practice.

Currently there seems to be a greater consistency between attitudes and behavior. If this is indeed the case, then the task of the researcher in quantifying, analyzing, and evaluating attitudes and behavior should be somewhat easier than it was in the past and may lead to a more valid assessment of beliefs, values, and actions.

Certainly, there is now an opportunity for extensive and thorough research on the effects of value changes, as they are manifest in public attitudes and actions relating to pornography, sex education, homosexuality, abortion, contraception, and other concomitants of change.

I suggest that there are many indications that the laboratory in human sexuality research is now as wide as the world. Proscriptions, prohibitions, controls, and fears have not disappeared, but, as is often the case, some segments of the public are ahead of the research community in expressing their perceived needs and in demanding more knowledge on which to base the new freedom of choice in sexual behavior.

As anyone who checks into a "singles bar" or frequents the gathering places of urban and suburban "swingers" knows, the apparent and overriding interest is in the *mechanics* of sex. As authors and publishers will attest, there seems to be no limit to the number of "how to" or "do it yourself" books that — to some extent, of course — play on the titillations of pornographic stimuli, but also appear to reflect another facet of the modern young person's concern with self.

In such a climate, we as research investigators and academicians are faced with a categorical imperative to go beyond the somewhat shallow considerations of popular mechanics and improve the quality and scope of knowledge of human sexuality and its significance in modern life.

Everyone here is and has been involved in this effort, but there has been very little coordination of studies and investigations among laboratories. Perhaps one of the reasons for this lack is the fact that there are no human sexuality centers that encompass the entire range of activities that are required.

Such a center should include a group of sophisticated researchers, a training program, a clinical center, and the administrative arrangements to allow for interaction among all of the center's components.

It is our plan at Stony Brook to develop such a center and to assemble a faculty and staff which represent a high research competence, clinical sophistication, and teaching and administrative excellence. Richard Green and Joseph LoPiccolo are now joining our faculty and we plan to bring additional experienced research and clinical people to our campus in succeeding years. Stony Brook will offer a unique opportunity for the development of such a center, because related departments within the entire University, such as psychology, sociology, and obstetrics and gynecology, are actively engaged in the field of human sexuality and all this work can be improved through vigorous interaction and administrative coordination.

This conference is, of itself, an indication that the development of such comprehensive centers is an idea whose time has come.

In providing funds for this meeting, the National Institute for Mental Health has asked us to concentrate on three specific objectives: first, to discuss and set forth the major priority areas for future emphasis in human sexuality research; second, to establish the basis for coordination in developing these priority areas; third, to concentrate on ways to test and demonstrate the possible linkage of human research and animal research in sexuality.

When we have achieved these objectives, the NIMH has asked us to prepare a statement of recommendations delineating the areas in which the Institute could best concentrate support on a priority basis. The significance of this request is extremely important. It indicates that research in sexuality — like research in drugs, narcotics, and alcohol — will be given greater support because of public pressures. Since its inception, the NIMH has supported research in all these fields, but in each instance only when controversy has spread throughout the population has the Federal government extended research support in efforts to solve problems of public concern.

None of us will argue the fact that there *is* public concern in this field. Public attitudes and public policies on topics such as homosexuality, pornography, abortion, and sex offenders have been the target of many different and strident pressure groups trying to impose their disparate beliefs on the public. But the facts — the knowledge — on which the public can base its decisions are neither adequate nor readily available.

Research on sexual behavior has been developing slowly, because of the difficulties of working in this area and because of the limited number of well-qualified researchers.

Research can, of course, be stimulated by an infusion of grant funds. But we have all learned in the past that, welcome as increased funding may be, it is wasted if the money is spent on "more of the same" or poured indiscriminately into untested programs whose results may be as harmful as the condition they were designed to control. In the drug field, the methadone program is considered by some to be a case in point. From its beginning, everyone involved *said* that distribution of methadone to heroin addicts would not solve the problem unless the addicts were given support in finding jobs and living drug-free lives. But only a few programs actually *did* much to follow through on what they knew was necessary.

In the field of human sexuality, we have an opportunity as well as a responsibility to shape the future. Research in physiology, dysfunction, and pathology must continue, but there is also a great need to relate biological factors to cultural patterns and to explore these biocultural issues.

There is also much benefit to be derived from an open and candid discussion of the various methodological problems which are particularly troublesome in the field of sex research. These range from problems of research design to issues relating to protection of the rights of human research subjects. And common to all our concerns is the need to coordinate our work, not merely to limit needless duplication, but to become mutually supportive.

I hope that there will be vigorous interaction among all participants on all subjects under consideration. All of us are in pursuit of excellence and the group is small enough to be exempt from the distractions of larger meetings.

And now, to work. Some day, we may know more precisely how human sexuality affects human behavior: in genetics, and in the physiological, psychological, and social bases of behavior. By the end of this conference, I hope we will know more precisely how we are to proceed as we extend and coordinate our investigations.

Early Sex Differences in the Human: Studies of Socioemotional Development[1]

Michael Lewis, Ph.D.[2]

In any discussion of sex differences one is bound to be confronted with the issue of the etiology of these differences. An implicit belief in the study of a particular individual characteristic such as sex role or gender identity is that the study of how early these differences might be observed would support one theoretical position or another. That is, if sex differences could be demonstrated in the very young, these differences would be due to biological or genetic determinants. It is not uncommon to find writers using early differences in sex role or gender identity as proof for the biological basis of their origin. Thus one of the first issues we must confront in looking at early differences is the issue of learning/experience vs. genetics.

Like many other issues in psychology, this dichotomy is rather useless since we are biological creatures, having sexual dimorphism, and we are also subject to wide and varying early learning experiences. From the outset it is therefore important to state that early sex differences in behavior do not necessarily reflect *either* experiential or genetic roots. Moreover, such questions usually do not add to our understanding of the phenomena.

Our research experience indicates that as early as we can find individual differences as a function of the sexual dimorphism in the infant so we can find a society expressing differential behavior toward the two sexes (Lewis, 1972a). For example, we (as well as others) have reported that girl infants at 12 weeks of age seem to be more attentive to auditory stimulation than boy infants (Lewis *et al.,* 1973). Thus from this early individual difference one might conclude that girls are biologically precocious, *vis a vis* boys, in auditory perception — which might account for their precocious language development throughout most of

[1] This paper was presented at the conference, "Sex Research: Future Directions," held at the State University of New York at Stony Brook, Stony Brook, New York, June 5-9, 1974.
[2] Institute for Research in Human Development, Educational Testing Service, Princeton, New Jersey 08540.

life. At the same time that one can demonstrate early sex differences in auditory perception, however, one can also find that mothers are responding differentially to their girl and boy children. Mothers of 12-week-old infants tend to speak more to those who are girls and, moreover, tend to respond more often to a girl child's vocalization behavior with their own vocalization (Lewis, 1972b; Lewis and Freedle, 1973). From the data it is not possible to determine whether the individual differences on the part of the infants are based on their differential experience via the mother or whether the mother's experience is determined by the child's differential response (biologically determined) to her auditory stimulation.

To repeat, we do not feel that the kind of question that asks us to differentiate the biological from the social will serve us in our exploration of early sex differences in socioemotional development. That is not to say that at some point in the future it may not be possible to untangle the biological-experience effects; however, at this time the expression of genotype in phenotypic behavior can be talked about only in terms of the interaction with the world, that is, with experience. Any attempt to separate them is probably fictional.

Neither in this paper nor in our research efforts do we attempt to discuss the variance accounted for by biology or experience. We recommend, therefore, that in general research not be undertaken with this distinction in mind. Thus the kind of research we are engaged in will not answer the question about the percent of variance accounted for by experience or biology, but rather seeks to explore the dynamics of the relationship and through an understanding of that relationship come to see whether behavior can be modified by experience (Lewis and Lee-Painter, 1974; Lewis and Rosenblum, 1974). The major issue is not the etiology of behavior — that is, whether it is due to biology or experience — but rather whether behavior is modifiable. Indeed, one could well argue that initial differences between the sexes may in fact be biologically determined, yet may be open to alteration by a wide variety of social experiences. It does not seem unreasonable, rather than to address ourselves to the basis of these early individual differences, to look toward the manner in which behavior might be altered.

In order to study the effects of experience, it is probably necessary to investigate an intervention paradigm wherein one tries to alter behavior via altering the environment. (Obviously, intervention of this sort has serious ethical consequences and needs to be approached carefully.)

If the study of early individual differences is not related to trying to separate out the variances accounted for by experience or biology, what use might the study of early differences have? The answer to this question really needs to be posed in terms of a more general question as to why study sex differences at all. In this context, a negative answer — that there is no reason to study sex differences — perhaps is more to the point. What I mean is that the study of socioemotional development can best be understood through the study of individual differences. Any individual variable which allows for variance in behavior will allow us to study process (Lewis, 1972c; Lewis and Wilson, 1972). Thus the study of individual differences is important inasmuch as it allows us to understand the process variables. In this particular case, by "process variables" we refer to the process of socioemotional development. In a conference on sex differences one point that I have chosen to make is that the study of sex differences needs to be secondary to the study of process. The reasons for this are both scientific and political. It raises the general reason for studying individual differences, and in this case sex differences *per se*. Why should anyone study individual differences as such? It seems to me that the study of individual differences without the study of process must be viewed in the context of an implicit set of assumptions about individual differences and the social consequences of these differences. Thus, rather than a truly scientific endeavor, it is a sociopolitical issue.

In this regard, the study of early differences in socioemotional development must proceed from the study of the socioemotional experience of children, their environment, and their interaction. Moreover, prior to experimentation an ethological approach must be undertaken. To fully appreciate the ethological knowledge, it may be necessary to initiate a series of experiments. The nature of these experiments is, of course, limited by important ethical consideration. What I have in mind, for example, is the observable fact that girl infants cry less than boys (Lewis, 1972b; Moss, 1967). Now this might have to do with the holding position of girl vs. boy infants. Thus it may be possible to show that the holding position, rather than any sex difference, is instrumental in producing the crying behavior and thus separate out the process from individual differences.

While a particular difference between male infant and female infant behavior may be *statistically* significant, this does not mean that the difference has any practical significance in the real world. Research concern should be concentrated on sex differences of practical significance rather than on those which are merely significant in the statistical sense of not being due to chance alone.

Our concern with male-female differences should not blind us to the fact that there are also wide differences among male infants and among female infants. Indeed, within-sex variability probably is as great as the between-sex variability.

RESEARCH EXAMPLES IN THE STUDY OF EARLY SEX DIFFERENCES

Experiential factors affect male-female differentiation almost from the moment of birth. Thus two studies, one in the United States and one in Britain, have reported differences in mothers' initial reactions during their first encounters with their babies, depending on the sex of the baby (Lewis and Als, 1975). For example, mothers feed boy babies more on the first day.

The study of behavioral dimorphism in infants can begin with simple ethological observations:

> Baby cries. Mother comforts baby. Baby stops crying but begins again. This time mother feels diaper, notes it is wet, carries baby to changing table — whereupon baby again stops crying.

We can then ask whether that observed chain of events varies with sex of the baby.

One useful probe for differences in parental and infant response is the study of parent-child interactions in families with opposite-sex twins. Studies of 18 such sets of twins have revealed a wide range of ways in which mothers treat male and female twins differently. Even if a mother does not dress a boy twin in pants and a girl twin in skirts, she is likely to select pants of appropriate color for each sex (Brooks and Lewis, 1974).

Still other experimental situations are possible which tap the young child's other social relationships as a function of sexual dimorphism: for example,

> Four mothers are seated in the corners of a square laboratory room, two holding male babies and two holding female babies. The mothers are instructed to place the babies on the floor facing the middle of the room. Under these conditions, the male babies are significantly more likely to crawl toward and play with each other rather than with a female baby, and female babies are also significantly more likely to crawl toward and play with each other than with a male baby. This is true for babies at twelve months of age as well as at eighteen months. It is true despite the fact that each male baby in this experimental setting faces two female babies but only one male baby (and *vice versa*); hence if the behavior were random one would expect crawling toward a baby of the *opposite* gender to be more frequent than crawling toward a baby of the same gender. The experiment also indicates that babies even at twelve months can distinguish the gender of other babies — though adults often find the distinction difficult to make. (Michalson *et al.*, 1974)

Observations to date indicate that a simple pattern of reward and reinforcement governs much behavioral dimorphism. Both parents and peers reward and thus reinforce those aspects of a child's behavior which are deemed appropriate to his or her gender; they reward negatively and thus inhibit gender-inappropriate behavior. Touching behavior, for example, is negatively reinforced for boys; this starts extremely early. Touching behavior is neither positively nor negatively reinforced for girls; essentially it is subjected to "benign neglect." Hence girls end up relatively freer to touch (Goldberg and Lewis, 1969; Messer and Lewis, 1972).

It is the mother rather than the father who controls the reward-reinforcement structure during the first year of life, at least in middle-class white American families. In one study, for example, parent-child interactions were recorded for 24-hr periods for babies at the age of 3 months. The fathers interacted with the babies for approximately 37 sec per day (Rebelsky and Hanks, 1971). In another study where families of several social classes were included, fathers were found to interact with their babies for an average of 15 min per day. The range was 0-2 hr. This study probably overstated the amount of interaction, since father-child interaction is socially approved behavior in this context, and fathers were aware that their interactions were being recorded. During the first year of a baby's life, in short, the pattern of nurture is matriarchal. During the second year of life, the father enters the situation. But he enters with male-appropriate behavior; that is, he *plays with* the child rather than tending it (Lewis and Weinraub, 1974).

In addition to biological and socioemotional factors, *cognitive* factors come into play only a little later, further modulating the process by which human behavior becomes dimorphic. Thus children by the age of 4 or 5 know perfectly well — in an explicit cognitive sense — whether they are male or female:

I am female. There are females and males out there. They behave differently. I will behave in the way appropriate to my femaleness.

The roots of this cognitive recognition of their own dimorphism go back at least to the twentieth to twenty-fourth month of life, when a child first begins to exploit the dimorphism of human language and to apply dimorphic labels — *mummy* vs. *daddy, him* vs. *her* (Michalson and Brooks, 1975). Precocious children may begin making the gender distinction even before the twentieth month; indeed, such gender labels may be the very first labels applied by the child. The ultimate adult human behavioral dimorphism thus has cognitive as well as biological and socioemotional roots (Lewis and Brooks, in press).

It seems to me that research in sex differences in socioemotional development must be based on the understanding of why we are interested in these individual differences. The study of individual differences is probably motivated by implicit variables having more to do with sociopolitical than scientific issues. Remember that for any important psychological variable there is greater within-group than across-group variance. The fact that we seek to concentrate on the across-group variance must reflect certain assumptions we make about the human condition. The study of sex differences in infancy is justified only inasmuch as it allows us to better understand the process at work in producing and affecting development.

Group Discussion

Dr. Rose questioned Dr. Lewis's belief that parental and infant factors in parent-infant reactions cannot be isolated. He pointed out that in ethological studies of monkeys it has been possible to distinguish clearly between sequences of mother-child interaction initiated by the child and sequences initiated by the mother. Dr. Green cited as an example studies showing that at a certain age male monkeys range farther away from their mothers than females — despite the more vigorous efforts of the mothers to retrieve the male infants. Hence it is not, in this case, greater maternal permissiveness which triggers the series of events, it is the male monkey infant who is unilaterally responsible. Dr. Lewis replied that this depends on where the series of observations begins. In the sequence of events described, the infant is clearly the initiator, but this does not mean that there were not maternal influences — perhaps the way in which the mother held the male child during previous weeks — which determined the wider ranging of the male offspring. At this stage in our understanding, it is better to describe the process than to seek to partial out responsibility. It is possible to give differing weights to the maternal and infant contributions during a given sequence of events even though it is not possible to separate them altogether (Lewis and Lee-Painter, 1974).

Dr. Lewis was asked about conflicting reports in the literature — reports of gender differences noted in some studies which could not be replicated. He replied that infant study procedures have not yet been sufficiently standardized to facilitate replication. To take an extreme case, the size and shape of a room may affect infant behavior; infants may respond in one way in a square room, differently in a corridor where their mobility is restricted. In one experiment with divergent-sex twins, one twin was tested with his or her mother while the other twin was kept waiting outside the test room. The usual female-male differences were noted with the twins first tested; that is, the girl twins showed more "proximal behavior" toward their mothers and more touching of their mothers, which the mothers accepted. When the second twins were tested, however, they had just suffered 15 minutes of separation — a stressing experience. The stressing did not alter the behavior of the girl twins, but the boy twins tested after the period of separation displayed more than the usual proximal behavior toward and touching of the mothers (Brooks and Lewis, 1974). Thus to replicate an earlier experiment one must not only replicate the actual test situation in full detail but also antecedent conditions and a wide variety of other, as yet not fully explored, variables.

Dr. Lewis was asked why, if intragroup variations among boys and among girls cover a broader range than intergroup differences between boys and girls, his work is in such large part concerned with intergroup differences. He replied that studying the differences between boys and girls is a useful probe in developing hypotheses concerning individual variations in general. By observing boys and girls separately, for example, he was able to note that girls spend more time in proximal relations with their mothers; this in turn may generate a hypothesis concerning the effects of proximal relationships on the developmental process in general. Male-female variations are interesting not only in themselves but also as clues to the determining factors in the developmental process.

Dr. Green pointed out that Dr. Lewis's data could be useful in another way. The data show that male infant behavior is scattered along a bell-shaped curve, and so is female infant behavior. A small percentage of females at one extreme of the bell-shaped curve are more malelike than the typical male, and a small percentage of males at the other end are more femalelike than the typical female. Longitudinal studies might then follow these atypical male and female infants through the years to determine the ultimate outcome of this early atypical behavior. Such longitudinal studies, Dr. Green pointed out, are his own field of interest. He therefore asked what particular behavior variables found in infantological and preschool studies might prove useful in identifying the particular atypical children whose subsequent development should be followed.

Dr. Lewis replied that while there were many differences between boy and girl infants he felt that they were less impressive than might be necessary to serve as the foundation of a longitudinal study. Further, it is not at all clear that a particular difference remains stable through time. For example, numerous studies show that male newborn infants cry more than female newborns (Moss, 1967). But studies also show that the boys who are crying most at 3 months of age are not the same boys who are crying most at 1 year. Indeed, there is a weak *negative* correlation between boys who cry a lot at 1 month and those who cry a lot at 1 year (Lewis, 1967). The reason appears to be that crying at 1 month is a response of vigor, instrumental in getting what an infant wants, while crying at one year is a passive response.

Asked about the next step in infantological studies, Dr. Lewis suggested that what was most needed in considering the relations between infants and their caretakers was to study just what "caretaking" means. To learn this, one must leave the laboratory, adopt ethological techniques, and actually observe how caretaking evolves in the home setting. How much in the course of an ordinary day do an infant and caretaker actually look at each other? How much is a child held? And so on. What are needed now are just data — excluding gender differences or anything else, simply determining what is in fact going on. What does the mother do on the first day she takes over care of the child — and what does the child do? What is happening on day 4?

Such studies are just beginning. Until the results are in, laboratory studies may be premature. For example, the laboratory can be used to study an infant's response to auditory stimulation. But what constitutes auditory stimulation for an infant? We must first observe the sounds to which he responds in a natural setting.

Sexual Identity: Research Strategies[1]

Richard Green, M.D.[2]

INTRODUCTION

Children from the first 2 years of life show an awareness that they are either male or female. Later they usually behave in ways culturally appropriate to this awareness. Still later they exhibit a preference for sexual partners of one or the other sex. These three developmental phases, (1) core-morphologic identity, (2) gender role behavior, and (3) sexual partner orientation, constitute *sexual identity*, a basic personality feature. My concern is with how sexual identity develops. Many strategies exist for answering this question; some will be reviewed here.

PRENATAL HORMONAL INFLUENCES

Do sex hormone levels before birth affect postnatal sex-typed behavior? Female rhesus monkeys exposed to large amounts of testosterone *in utero* behave more like young male monkeys than female rhesus monkeys whose prenatal hormonal environment has not been altered; they become "tomboy" monkeys (Young *et al.*, 1964). Postnatal exposure to this same androgen does not have a comparable behavioral effect.

Is this phenomenon also true of humans? While the hormonal exposure of the human fetus cannot be altered experimentally, a genetic defect has performed this "experiment" for us. In the adrenogenital syndrome, the female fetus produces excessive adrenal androgen. Girls with this syndrome, like female rheus monkeys also exposed to high androgen levels *in utero*, appear more

[1] This paper was presented at the conference, "Sex Research: Future Directions," held at the State University of New York at Stony Brook, Stony Brook, New York, June 5-9, 1974.
[2] Department of Psychiatry and Behavioral Science and Department of Psychology, State University of New York at Stony Brook, Stony Brook, New York 11794.

tomboyish than typical girls. When compared with their hormone-normal sisters, they are *less* interested in doll play and wearing dresses and are *more* interested in rough-and-tumble play (Ehrhardt *et al.*, 1968a; Ehrhardt, 1973). However, the effect of these hormones on adult sexual identity is not as dramatic; studies of adult women with the adrenogenital syndrome do not reveal an excess likelihood of transsexual or homosexual behavior (Ehrhardt *et al.*, 1968b).

There is an analogous phenomenon in human males. For about 20 years, it was the practice in one clinic to give pregnant diabetic women large doses of estrogen and lesser amounts of progesterone to compensate for a suspected hormone deficiency complicating the pregnancy of diabetics. Comparing two groups of sons, aged 6 and 16, born to these women with the sons of non-hormone-treated women, we found that boys exposed prenatally to exogenous female hormones were *less* rough-and-tumble, *less* aggressive, and *less* athletic. The possibility remains, of course, that the behavioral differences were related to the mother's chronic illness rather than to altered hormonal levels *in utero* (Yalom *et al.*, 1973).

In a section below, boys who are behaviorally very feminine and who also do not engage in rough-and-tumble play are discussed. It should be noted here, however, that the boys born to the diabetic mothers, although less rough-and-tumble than the contrast group, were not overtly "feminine." Thus, while it may be that an innate low level of aggressivity contributes to feminine behavior in boys, a particular style of parenting (Stoller, 1968; Green, 1974), perhaps one which positively reinforces feminine behavior, may *also* be required for this innate contribution to permit the full picture of boyhood femininity to develop.

Research must distinguish direct from indirect effects of prenatal hormone exposure. In a culture which does not label some forms of childhood behavior (such as rough-and-tumble play) as masculine and others as feminine, subtle effects of prenatal hormone exposure might pass unnoticed in the welter of other individual differences. In our culture, however, the same variations may have far-reaching consequences. A boy with less than the usual boyhood aggressivity may be labeled "sissy" — with profound effects on peer-group socialization and father-son interaction. A low level of aggressivity may result in a boy's accommodating more easily to the company of women and girls. "Boys play too rough" is an explanation given by the feminine boys described later who prefer female companionship and have an alienated relationship with father.

How can we study the effects of prenatal sex steroid levels in the normal human? While we cannot experimentally alter the prenatal hormonal milieu, we may be able to measure it — perhaps by periodically sampling amniotic fluid or by periodically assaying maternal plasma or urine. If a reliable index of the hormonal exposure of the human fetus could thus be found, we might begin to trace correlations between prenatal hormones and neonatal and childhood behaviors. We might also be able to determine the stages of fetal development — critical periods — during which the hormone exerts its later behavioral effect. Studies such as this, despite their science-fiction flavor today, should not be excluded from an agenda for future research.

EARLY-LIFE SEX DIFFERENCES IN BEHAVIOR

Neonatal and infant studies designed to assess sex differences have been described by Lewis; therefore, only brief attention will be paid to this research strategy. I endorse the premise that studies of the human neonate hold considerable potential for isolating the early roots of male-female dimorphism. These reported sex differences which group into displays of greater muscle strength, sensory differences, degree of affiliative behavior to adults, and patterns of maternal care have been reviewed elsewhere (Green, 1974; Maccoby and Jacklin, 1974a).

I will here underscore a strategy in utilizing these differences — that of studying atypical children of both sexes. Many measures have a bell-shaped distribution. Males and females who fall at the ends of the distribution could be longitudinally studied to determine correlations between neonatal behavior and later developmental attributes. Of interest would be those infants whose patterns fall within the zone typically found for the other sex. For example, will males with a female pattern of taste preference (greater appeal of sweet) or within the female range for elevating the prone head (less ability) show later childhood behaviors which are culturally feminine, e.g., preferring doll play to rough-and-tumble play? Will infants handled by mothers in a manner more typical for other-sex infants show later atypical sex-typed behavior?

ANATOMICALLY AMBIGUOUS CHILDREN

Next consider infants with anatomically ambiguous external genitals (pseudohermaphrodites). Among infants in whom the ambiguity is great, some have been labeled male at birth and raised as boys; others with the same genital ambiguity have been labeled female at birth and raised as girls. Both groups generally develop a sexual identity consonant with the sex to which they have been assigned and as which they have been consistently reared (Money *et al.*, 1955; Stoller, 1968; Money and Ehrhardt, 1972).

An infant with the virilizing adrenogenital syndrome who is designated male at birth and raised male, despite a female chromosome array (44 plus XX), and despite the presence of ovaries and a uterus, will typically identify as a male, show masculine behavior in childhood, and be erotically attracted to females in adulthood. If designated female and raised as female, the child will identify as female, show primarily feminine behavior, and later be erotically attracted to males. Such "experiments of nature vs. nurture" indicate that environmental factors can outweigh whatever innate biological influences may be present in such cases.

From these patients we learn more. They teach us when the earliest components of sexual identity are "set." If such an infant is unambiguously raised as a male for the first few years (up to approximately the fourth birthday), subsequent reassignment as a female is generally not successful (Money et al., 1955; Stoller, 1968).

These classical studies and concepts, first announced nearly 20 years ago, have been challenged (Zuger, 1970; Diamond, 1965). More recently, they have also been confirmed — even in cases where sex of assignment is in dramatic conflict with genital appearance. Thus an adult female with the adrenogenital syndrome, raised as a girl, is described as feminine and heterosexual despite the presence of a 7-cm clitoris, her major obstacle to heterosexual coitus (Lev-Ran, 1974).

Some investigators have challenged the model of the anatomically intersexed child as appropriate for the study of "typical" sexual identity development. However, a study of monozygotic male twins currently under way may answer some of these objections. As a result of early-life circumcision accident, one twin suffered a sloughing of the penis and is being raised as a female. That child appears to be developing a female identity (Money and Ehrhardt, 1972).

Studies of monozygotic twins offer additional promise. Our research has identified two sets of twins, one set male and one female, with discordant sexual identities. One male twin at the age of 10 is very feminine and wishes to become a girl; the other is typically masculine. The 25-year-old female pair consists of one person who urgently seeks sex-change surgery and a feminine sister. Differential early-life socialization experiences are described for both twin pairs (Green and Stoller, 1971). In such twin studies, prenatal hormonal and genetic differences can be given less consideration as an explanation for behavioral differences.

Do adults with "intersexed" sex chromosome abnormalities — such as the XXY configuration — also show "intersexed" behavioral differences? Some studies indicate that they do (Money and Pollitt, 1964; Baker and Stoller, 1968). Our difficulty is that we know only those cases which are reported. We do not know what proportion of XXY or XYY children and adults are leading lives without conspicuous behavioral sequelae such as transsexualism or transvestism. Here is another research opportunity. One male infant in 700 is born with the XXY chromosomal pattern. These infants look normal at birth but can be identified by karyotyping. Thus, if we karyotype an extensive series of newborn male infants, we can expect to identify about 14 infants with this syndrome out of 10,000. Followups on these 14 would provide a study of the effects of this atypical karyotype on subsequent behavior in a random sample. Large-scale karyotyping studies are in fact currently under way.

Study of neonatal sex differences could also find applicability with samples of the anatomically intersexed. If one or another sex difference is replicated on normal infants (e.g., taste preference), would anatomically intersexed babies (e.g., adrenogenital females or XXY males) also be behaviorally intersexed?

Neonatal karyotyping studies, however, raise eithical questions. If an infant is identified as XXY, should this information be withheld from or provided to parents? If parents receive input, how will this affect the child's rearing? Would any possible adverse effects outweigh potential benefits? One civil suit in Maryland recently blocked the identification of XXY male children on the grounds that such identification may stigmatize.

CHILDREN OF THE ATYPICAL

Increasing numbers of children are being brought up by homosexual couples or in a home where one parent has undergone sex-change surgery. Transsexual couples are adopting children neonatally; others are rearing children born to the prior marriage of one partner. Females married to men who were formerly women are giving birth to and raising children conceived through donor insemination. In a recent court case, a chromosomal female now living as a man was awarded custody of a child of an earlier marriage; "he" is thus now in the role of father of the child of whom he was formerly the mother.

Lesbian mothers and couples are receiving considerable attention in the press and in courts of law, where they are fighting for child custody. What effect does being raised by one or two homosexual adults have on children? What is the effect on a child of recognizing that a parent is homosexual, or of peer group reactions to a child's atypical household?

Role models of the other sex and the model of heterosexuality are not omitted from the lives of children who live in homosexual-parent households. Children are repeatedly exposed to heterosexual adults of both sexes in the person of relatives, parents of peers, and adults at school. Then there is the additional exposure to the conventional nuclear-family constellation on television and in books.

The effect on a child's later sexual preferences of knowing that at least one parent is homosexual is questionable. This will partly depend on the degree to which partner preference is a result of role modeling. Role modeling cannot account for the *entire* process of sexual identity development, however, in that the vast majority of homosexuals were raised by *heterosexual* parents. The view held by the homosexual parent, or couple, of individuals of the other sex may be significant. The image, positive or negative, painted of these absentee figures may shape later affectional elements.

Yet another issue is a possible biological predisposition to homosexuality, perhaps inherited, such as atypical gonadal hormone levels. Should such a developmental basis be demonstrated, would raising a child so predisposed in either a homosexual or a heterosexual household significantly affect future sexuality?

These "social experiments" will teach us much about the development of sexual identity and are yet to be harnessed (Green, 1975).

THE CHILDHOOD OF ATYPICAL ADULTS

The first strategy to be adopted in the effort to unravel the etiology of adult sexual atypicality was that of Freud asking atypical adults about their childhood. This has been the traditional approach to understanding the parents of the atypical adult and his or her early socialization experiences. This technique, although limited, has research value.

Case histories given by transsexuals reveal that there are anatomically normal adults with an intense, irreversible conviction of belonging to the other sex who recall the onset of this cross-sex identity during early childhood (Benjamin, 1966; Stoller, 1968; Green and Money, 1969). Invariably, these persons report having role-played as persons of the other sex, having dressed as children of the other sex, having preferred opposite-sex children as playmates, and having avoided the toys and games typical for their sex.

Retrospective histories of transvestites (heterosexual males who cross-dress with accompanying sexual arousal and do not desire sex change) also demonstrate the early-life onset of atypical sexuality. Approximately half of 500 transvestites in one series (Prince and Bentler, 1972) reported commencing cross-dressing prior to adolescence.

Retrospective reports by homosexuals again point to the enduring significance of atypical childhood gender-role behavior. One study reported that about a third of 100 homosexual adult male patients recalled playing predominantly with girls during boyhood (compared to 10% of the heterosexual control group) and 83% displayed an aversion to competitive group games (compared to 37% of the heterosexuals, Bieber *et al.*, 1962). Another study of a nonpatient homosexual sample reported that two-thirds of 89 males recalled "girl-like" behavior during childhood (compared to only 3% of the heterosexual controls). For female homosexuals, over two-thirds of a group of 57 were tomboyish during childhood (compared to 16% of the heterosexuals), with half persisting with tomboyism into adolescence (compared to none of the heterosexuals, Saghir and Robins, 1973).

But retrospective approaches have shortcomings. First, we cannot be sure how much the reports are contaminated by knowledge of the outcome. They are certainly not made up out of whole cloth, however; family picture albums, for example, sometimes show a male transsexual or transvestite already dressing in women's clothes and struggling with high-heel shoes at the age of 2 or 3 (Green, 1974; Stoller, 1968). But the detailed sequence of events in the early years is not documentable in this way. Prospective studies, beginning at an early age, are needed to circumvent these shortcomings. Further, retrospective studies cannot tell us how many children who display atypical sex role behavior later "outgrow" it and become adults without an atypical identity.

Prospective studies are not without *their* problems, however. A percentage of subjects are always lost at follow-up. And (unless a large control group is included) they may cast little light on the large number of behaviorally typical boys who accept themselves as male, who display masculine behavior, but who in adulthood display a preference for male sexual partners.

ATYPICAL BEHAVIOR IN ANATOMICALLY NORMAL CHILDREN

My principal research strategy consists of identifying, at an early age, a population of anatomically normal children who display atypical sex role behavior. These children are matched with a control group of children displaying typical sex role behavior. Both groups are then followed longitudinally.

We have generated a sample of 65 boys, under the age of 11, who show a strong preference for the activities, toys, dress, and companionship of females, state that they want to be girls, and role-play typically as females in "house" or "mother-father" games. These children are being compared with typical boys of the same age, sibling order, socioeconomic level, and family constellation. So far, we have matched 50 boys. While we are necessarily dependent on retrospective inquiry in reconstructing the sequence of events before a child enters the study, recollections when a child is 6 or 7 should be less subject to error than after another two decades.

Parents and siblings of both groups are studied via a variety of procedures (Green, 1974). The boys referred because of "feminine" behavior test similarly to same-aged girls (another control group) on a variety of psychological procedures and differently from control boys. When constructing fantasies, they typically utilize female family figures and an infant (as do girls, whereas boys utilize male figures and pay much less attention to an infant). When requested to draw a person, the figure drawn is usually female (girls do the same; most boys draw a male). Left alone in a playroom stocked with sex-typed toys, they play mostly with a "Barbie" doll (as do girls, while other boys play with a truck). On the It-Scale for Children (in which a "neuter" figure "It" selects a variety of sex-typed preferences illustrated on cards, Brown, 1956), their selection for sex-typed toys, playmates, and accessories is the same as that of girls but differs from that of most boys. When they complete picture card sequences in which a child of their own sex joins a parent engaged in an adult sex-typed activity, they join the female parent in a feminine activity (as do girls, but not most boys, Green, 1974). Additionally, preliminary spectral analyses of electroencephalographic recordings on a subsample of boys from both groups provide patterns which correctly discriminate the feminine from masculine boys with 90% accuracy (Hanley and Green, unpublished).

In every family with one behaviorally feminine boy there is not a *second* behaviorally feminine boy. Why? Our attention must also focus on within-family differences, not merely on ways in which feminine-boy families differ from control families. Features which distinguished the feminine boy from siblings during the first years and styles in which parents differentially responded to the femininely behaving boy receive close scrutiny.

A tentative review of findings suggests that this feminine behavior evolves as an interactional, sequential effect engaging innate features of the child and early postnatal socialization experiences (Green, 1974). These children may indeed be innately less rough-and-tumble and aggressive. This translates into differential mother-child and father-child interactional patterns, and further affects the child's early peer socialization. These boys typically complain that "boys play too rough." Their early socialization experiences are usually with girls, and greater competence in feminine socialization patterns develops during the preschool years. Feminine interests help promote a closer affective relation with their mothers but contribute to alienation from their fathers. Not getting the affirmative feedback that he expects or that he gets from his other son(s), the father labels the boy a "mama's boy." During early school years, male peer group alienation evolves. The boy is labeled "sissy." Ostracism escalates as the rejected boy gravitates more toward the female group and begins to show feminine gestures which "tag" his uniqueness and isolate him further from male peers.

The effect of same-aged peer group relations during grade school years deserves more study. Boys with girl "friends" during boyhood are more likely to have male "lovers" during adulthood. The relationship between early peer group interaction and later genital sexuality is enigmatic. One possibility is that the feminine boy's lack of positive affective responses from males during earlier years (peer group and father) yields "male affect starvation," compensated for in adulthood by male-male romantic relationships. Alternatively, the young male whose peer group is female may be so completely socialized within that group that he evolves similar later romantic attachments (males). The manner in which preadolescent *homosocial* peer group relationships typically evolve into adolescent and adult *heterosexual* relationships and *heterosocial* peer group relationships into *homosexual* ones is a little understood facet of sexual identity development. These very active "latency" years deserve more study.

TREATMENT OF THE CHILD WITH ATYPICAL
SEXUAL IDENTITY

Intervention into the behavior of the very feminine boy engages both research and ethical issues. First: research questions.

What do we know about the natural course of boyhood femininity? What might intervention do? Follow-up studies tell us something of the natural course. Twenty-seven adult males previously seen for boyhood femininity have been reevaluated in three studies. Fifteen are currently transsexual, transvestic, or homosexual (Lebovitz, 1972; Zuger, 1966; Green and Money, cited in Green, 1974). The manner (if any) of treatment intervention with most of these patients is not clear. By contrast, the adult transsexuals, transvestites, and homosexuals in the series noted earlier were rarely *evaluated* during childhood, let alone treated.

Consider again the three components of sexual identity: earliest, self-awareness of being male or female (core-morphologic identity); later, culturally defined masculine and feminine (gender-role) behavior; still later, preferred sex of partner for genital sexuality. All three components are atypical for male-to-female transsexuals (they consider themselves female, behave like women, and are attracted to anatomically same-sexed partners). Transsexuals report that their parents felt their atypical behavior was insignificant and would pass, and so they were not treated. For them, sexual identity remained fully atypical. Although a few boys seen by Money and myself over a decade ago were initially behaving similarly to that recalled by transsexuals, they underwent evaluation and are not now transsexual. Most are homosexual. Why did the first two components of sexual identity undergo change and the third component remain atypical?

We can speculate. Core-morphologic identity appears to be set to a substantial degree during the first 2-3 years of life. Thus a considerable portion of sexual identity has evolved by the time the atypical child is initially evaluated. Gender-role behavior, a later identity component, may still be more modifiable, however. Therefore, one outcome of early intervention may be that a young male who feels he is or wants to be a female might be convinced that the change is not possible and, with a change in his milieu (pleasurable nonfeminine activities and non-rough-and-tumble male playmates), may experience adequate comfort with anatomical maleness and male peer group socialization. If so, the later request for sex change may be averted. However, no attention is being specifically addressed to the third component, genital sexuality, since such behavior is not manifested during these years.

The critical factor at issue in whether the behavior of an atypical boy changes may be whether the parents seek evaluation. Those parents who request evaluation are initiating a new milieu for their son, one which discourages feminine behavior and encourages masculinity. Because of this, the second component of sexual identity, and perhaps indirectly the first component, may modify. If such is the case, the pre-transsexual male may then mature into a homosexual male. (The degree to which change in the first two components might influence the third component is even less clear.)

Should clinicians attempt to modify the behavior of the child whose sexual identity is significantly atypical? The very feminine male child experiences considerable social conflict in consequence of his behavior. He is teased, ostracized, and bullied. Parents who bring their very feminine boy for professional consultation are concerned about his behavior and want something done. What then is the professional's responsibility toward the parents and their child?

It can be argued that the conflict experienced by the feminine boy derives mainly from the culture in which he lives, a culture that dictates, for irrational reasons, that boys and girls behave in specified dimorphic ways. While some societal change is taking place, most children continue to label feminine boys "sissy." Unless the entire society undergoes dramatic ultrarapid change, the distress experienced by today's very feminine boy will augment during teenage. While the clinician may prefer that the pediatric and adult culture immediately change, there is more basis for optimism in helping a single individual to change.

But what kind of change? Treatment need not forge the feminine boy into an unduly aggressive, insensitive male. However, it can impart greater balance to a child's interests and behavior where previously skewed patterns have precluded comfortable social integration. For example, consider the exclusively female peer group of the feminine boy. Intervention may help the feminine boy find unstigmatized boys who prefer "sex-role-neutral" activities, thus widening his range of social interactions.

Clearly, "to treat or not to treat" is a dilemma. The standard of many clinicians is to be nonjudgmental, with the patient dictating goals. In the event that (1) parents want their child to be happier, (2) the child is in serious conflict, and (3) the likelihood of reducing that conflict is greatest by promoting behavioral change, is it ethical to refuse intervention? A more extended discussion of this ethical quicksand is found in Green (1974, 1975).

TOMBOYISM AND ADULT SEXUAL IDENTITY

Tomboyism is much more common than its counterpart, boyhood femininity. However, adult females who request sex-change surgery and who typically recall tomboy behavior during childhood are outnumbered 3:1 by males requesting sex change. Although two-thirds of a recent sample of adult female homosexuals also recall having been tomboys (Saghir and Robins, 1973), the incidence of female homosexuality appears to be about half that of male homosexuality (Kinsey *et al.*, 1948, 1953). Thus tomboyism can, in more cases than boyhood femininity, be validly considered a "passing phase" with no adult counterpart in atypical sexual identity.

Yet there do exist some tomboys for whom this cross-gender *behavior* reflects a fundamental feature of *identity*, a feature which will continue beyond childhood and express itself in transsexualism or homosexuality. But, in contrast to very feminine behavior in boys, masculine behavior in girls causes less concern on the part of parents and less conflict for the child. Thus clinical facilities are less likely to have tomboys referred for evaluation. Consequently, we have insufficient data to determine which few tomboys are those whose sexual identity will be atypical during subsequent years, although some pilot data have been reported (Green, 1974).

SEXUAL IDENTITY AND GONADAL HORMONES

Following decades during which a hormonal basis for atypical human sexual behavior fell into disrepute, a revival of interest exists. Exquisitely sensitive hormonal assays have opened a new era of investigation. Where previous studies utilizing gross, nonspecific measures failed to show differences between homosexuals and heterosexuals, some recent studies have found differences. The studies are several and have been compiled elsewhere (Green, 1975).

More important than reciting the provocative, although conflicting findings are the questions of specificity regarding the reported differences. Rigorous attention has not been paid to possibly confounding variables, especially stress, drug intake, and recency of sexual activity. Heterosexual males under military stress have testosterone levels lowered to the same degree as those reported for some homosexuals (Rose *et al.*, 1969; Kruez *et al.*, 1972). (Homosexuals, because of their stigmatized life style, may be under greater stress than heterosexuals.) Data on marijuana ingestion may be given for one group of subjects (homosexuals), but not for the other. Tetrahydrocannabinol appears to reduce plasma testosterone (Kolodny *et al.*, 1974). Whereas one study (in contrast to the remainder) found homosexuals to have higher androgen levels (Brodie *et al.*, 1974), data are not given for sexual activity prior to plasma sampling. (Sexual activity can raise plasma testosterone: Fox *et al.*, 1972; Pirke *et al.*, 1974.)

Individual variations in both levels of plasma testosterone and sexual activity range widely for the human. Also, the correlation between the two is not very high; indeed, some castrated males retain sexual activity for years. Thus individual or group differences may have little behavioral meaning. Additionally, the range of individual tissue responsivity to the same level of hormone may vary widely, making control for tissue response possibly even more significant than controlling for the other confounding factors noted above.

An alternative strategy is collecting longitudinal assays on children with atypical behavior from childhood through adolescence and into adulthood. Correlating developmental behavioral attributes with hormonal maturation could constitute a significant integration of the neuroendocrine and developmental psychology disciplines and promote a new research dimension.

TYPICAL CHILDREN WITH ATYPICAL BACKGROUNDS

In the event that certain variables appear to predispose a child to a later same-sex sexual partner preference (for example, father-absence in the male, to cite a common theory), then children without fathers who do *not* become feminine or homosexual could be studied to determine the roots of their heterosexuality. A too infrequently used strategy is study of the "high-risk" child who develops normally to ascertain "what went right."

BISEXUALITY

The term "bisexuality" (or "ambisexuality") has many usages. Here it is narrowly restricted to those persons who rate "3" on the 7-point Kinsey scale, with 0 designating exclusive heterosexuality and 6 exclusive homosexuality. Individuals equally disposed in fantasy and overt behavior to males and females are not common.

True bisexuality raises theoretical and research questions. Explaining homosexuality as an anxiety or phobic reaction to one genital configuration (typically the male rendered anxious by the "castrated" female) is problematic in understanding the individual capable of sexual satisfaction with both sexes. Further, if future research documents that specific developmental routes, social or biological, promote either an exclusive male or an exclusive female sexual partner preference, would bisexuals fall in the middle range for these attributes? Finally, will the current changes in early childhood socialization that less clearly demarcate sex roles promote more bisexuality during adulthood?

CROSS-CULTURAL STUDIES

Studies of atypical children need to be conducted in cultures with different patterns of child-rearing and different definitions of "masculinity" and "femininity." The anthropological literature is replete with examples of adults and children who have adopted the sex role opposite to that expected by virtue of their anatomy (Green, 1974). Missing from these accounts, however, is *why* these atypical individuals adopted their unique life style.

CONCLUSION

These strategies are varied. They are ambitious. Pieced together, addressed by scholars from developmental psychology, neuroendocrinology, human ethology, anthropology, and sociology, the puzzle will reveal a most fundamental behavioral attribute. The designs are available. The tools are forged. The researchers are trained. The populations wait. Will the support be forthcoming?

Group Discussion

Dr. Cole pointed out that a study of the physically handicapped groups with whom he is concerned might throw light on the development of gender role and gender identity. Society treats some categories of the handicapped as if they had no genitals at all; the existence of genitals is completely ignored. What are the sequelae of this kind of treatment?

Dr. Green asked about studies of the congenitally blind. For example, what are their early masturbatory fantasies? Dr. Green was asked, with respect to certain photographs of patients which were projected during his talk, whether he had secured the permission of the patients to display the photographs before scientific audiences. He replied that he had.

Dr. Green was also asked whether, among homosexual males, there were any childhood differences in sexual behavior between those who were feminine in dress and manner and those who were more masculine in these respects. Dr. Green replied that he knew of one study conducted in Mexico indicating that those who played with dolls as children and in general were identified as "sissies" played the "passive," "insertee" role in adult homosexual relations. Those who engaged in rough-and-tumble play, and who in other respects were unremarkable as boys, played the "active," "insertor" role (Carrier, 1971).

Dr. Green was asked whether the distribution of male and female siblings in the nuclear family may have influenced the development of effeminacy in the boys he was studying. He replied that the sex of siblings of his patients (and thus of his control group) is, so far, randomly distributed.

Dr. Green was asked to describe more how the type of mothering a child receives is being studied. He replied that he questions mothers and fathers in detail, in a semistructured interview, concerning their early impressions of the patient as an infant, as compared with their son's siblings, how much the patient and each of his siblings were held, how each reacted, and so on, year by year. Responses are tape-recorded, transcribed, and blind-rated by research assistants.

Dr. Schmidt referred to Dr. Green's suggestion that the greater societal pressure on boys to conform to male role stereotypes might explain why more boys than girls seek sex change in adult life. He asked whether direct evidence for this could be secured. If so, it might be of great practical importance in reducing the incidence of dissonant gender identity.

Dr. Green replied that such evidence might soon be forthcoming. In some subcultures recently, the pressure on boys to conform to the male stereotype has been greatly weakened. As a result, boys who are not athletic and who shun rough-and-tumble play may no longer be extruded from their male peer group and may no longer be forced to gravitate to the female group. If this weakening of pressures to conform does in fact result in fewer boys from such subcultures appearing at sex-change clinics, the hypothesis will be supported. Similarly, it will be interesting to note whether recent changes in peer group pressures in childhood produce a change in the incidence of homosexuality. All we can say now is that there are many boys today who, at the age of 5, 6, or 7, insist that they want to be girls because "boys play too rough." Since they see things in black-and-white terms at that age, being girls seems to them the only logical alternative to being roughed up. It may be that if sex roles continue to merge, those children who now feel driven by peer pressures will feel less driven; but there will still remain a core group of children who, *before* the rise of peer pressures, are already identifying with the other sex. When a child at 13 months of age is walking in high-heeled shoes, and at 18 and 23 months of age is role-playing exclusively as a female, his behavior is probably not attributable to peer group pressure. Hence we may have two groups of children displaying atypical sexual identity — those for whom changing peer group expectancies will have greater or lesser influence.

Dr. Green was asked what would happen if his data on the ways in which a child's gender role behavior is influenced by environmental factors became generally known to parents — or, in a more extreme case, what would happen if parents were trained to follow patterns which discourage gender role dissonance. Would this result in fewer dissonant children? He replied that he doubted it. One earlier study compared parental expectations of male and female behavior in their children with the behavior actually observed in a nursery school setting. There was no correlation (Sears *et al.*, 1965), either because parental wishes and plans are not translated into effective parental action or because a child's gender behavior is child-activated, innate, or otherwise not responsive to the expectations of parents.

Dr. Lewis pointed out another possible explanation. Parents might report and might believe that they were encouraging masculine behavior in their sons despite the fact that their actual approach was encouraging feminine behavior.

The question was raised whether sexual anxiety in the parents might be a factor in dissonant sexual development in the child. Dr. Green replied that he was asking both parents of feminine and masculine male children questions related to sexual anxiety — attitudes toward masturbation, sex play, household nudity, and so on. The data are on audio tape and will be analyzed.

Dr. Lipman-Blumen underscored the point that the larger number of boys brought to clinics for effeminate behavior as compared with girls brought for tomboy behavior might be a poor guide to the actual incidence of effeminacy in boys and tomboyism in girls. Perhaps, since male behavior is so much more highly valued in our culture, the parents of tomboy girls are less likely to bring them to a clinic than the parents of effeminate boys. Dr. Green agreed and said that moderate tomboyism was probably more common than moderate feminine behavior but that significant effeminacy might be more common than significant tomboyism.

Dr. Rose noted that some earlier studies had reported a high incidence of "absent fathers" in the case histories of homosexuals, while other studies had found a normal incidence. Dr. Green replied that in his clinical sample (a predominantly white sample), about 40% of the children had been permanently separated from their biological fathers by the age of 4 — a much higher rate than in the general population (census data). However, the question remains whether the same would hold true for another clinic (non-gender-problem) population. This could be checked, Dr. Green noted, by comparing absence of the father among children brought to a clinic for gender role problems with that among children brought for other reasons, such as enuresis. Dr. Green pointed out that families with one dissonant child rarely have two — so it cannot be merely the absent father which dictates the outcome. Something must be uniquely happening to the atypical child.

Dr. Bell stated that a wide difference should not be expected between children with gender problems brought to clinics and those with gender problems not brought to clinics. His impression of his own data on large numbers of homosexuals indicates a relatively modest difference between clinic and non-clinic populations.

Dr. James pointed out that anthropologists might study populations where the fathers are routinely absent 9 months or more a year for work reasons.

Dr. Rose called attention again to a potential flaw in retrospective studies. Patients often read the scientific literature, and perhaps parents do, too. May it not be that when they are later questioned they play back the syndromes and surrounding data which they have found in the literature? Should they not be asked about their reading? Dr. Green replied that transsexuals in particular have read widely in the psychiatric, psychological, and popular literature, and this does in fact confound their retrospective reports. Dr. Cole noted a similar phenomenon with handicapped children; the parents read about the handicap, then come in and report the symptoms they know from their reading they should be reporting.

Dr. Rose also called attention to the fact that the desire to please may affect retrospective responses. Thus a respondent who volunteers for a study may be motivated to give the responses he thinks will please the inquirer. This may explain, for example, why a homosexual gives answers which the inquirer expects with respect to his early relations with his parents — even though he is politically hostile to the view that his homosexuality stemmed from such factors.

Dr. Green was asked about parental strife or discord as a factor in the development of his patients. He replied that this was being studied, both through questionnaires and through bringing parents together for laboratory reconciliation of provoked discord (the "revealed differences" technique).

Dr. Gebhard recalled an occasion when he asked Dr. Hooker why males seemed to have almost a monopoly on certain forms of sexual deviance — fetishism, sadomasochism, and certain other paraphilias. Perhaps, Dr. Hooker had replied, it is because women bring up both girls and boys. Girls can thus follow the maternal track, while boys must break from that track and find tracks of their own. Perhaps if fathers raised daughters we would have a full range of female paraphilias, too.

Male-Female Differences in Sexual Arousal and Behavior During and After Exposure to Sexually Explicit Stimuli[1]

Gunter Schmidt, Ph.D.[2]

INTRODUCTION: THE CONVERGENCE OF MALE-FEMALE SEXUAL BEHAVIOR PATTERNS

The latest survey data on male-female differences in sexual behavior in West Germany show two trends (Schmidt and Sigusch, 1971; Sigusch and Schmidt, 1973). Today's adolescents and young adults have a sex-specific view of sexuality which is clearly related to the traditional sexual gender roles: girls behave sexually as if they had less sexual drive than boys; girls show fewer signs of sexual frustration when they abstain sexually; girls behave as if they should show less sexual initiative than their male partner or at least not more than he; girls behave as if their sexuality is more dependent on love, personal relations, and fidelity than do boys. The stereotype of the less libidinous, less initiative-taking woman whose sexuality can be realized only within emotional and personal relations is still of central importance.

During recent decades, however, and especially during the last 10 or 15 years, these differences have decreased considerably. There is a convergent trend resulting in a steady decline of the sex differences described above. To mention only a few findings: the overwhelming majority (about 90% of both young women and men) now have permissive and egalitarian sexual standards; the incidence of premarital sociosexual activities (necking, petting) and of premarital coitus for women and men is equally high; girls have the first coital experience at nearly the same age as boys (about half a year later). The gender differences

[1] This paper was presented at the conference, "Sex Research: Future Directions," held at the State University of New York at Stony Brook, Stony Brook, New York, June 5-9, 1974.
[2] Institute for Sex Research, University Psychiatric Clinic, University of Hamburg, Hamburg, West Germany.

regarding partner multiplicity have converged; on the one hand, the high degree of partner mobility among men, which resulted from the extreme double standard, has decreased; on the other, women now show a greater tendency not to limit their sexuality to one partner. The differences regarding the age of first masturbation and the accumulative incidence of masturbation have decreased considerably; however, masturbation remains the type of sexual behavior with the most distinctive sex differences. This convergence process can be observed in all Western societies for which we have data (Zetterberg, 1969; Israel *et al.*, 1970; Christensen and Gregg, 1970; Christensen, 1971; Bell and Chaskes, 1970; Sorensen, 1973; Hunt, 1974). It seems to be most advanced in the Northwest European countries (Denmark, Sweden, West Germany) and in the metropolitan areas along the East and West Coasts of the United States.

This paper will deal in detail with one symptom of the convergence process: the decreased differences between the sexes in the reactions to pictorial and narrative stimuli. Kinsey *et al.* (1953) gave evidence of considerable differences in the responsivity of the average male and female to pictorial and narrative sexual stimuli. They found that arousal from explicitly sexual stimuli was much rarer among women. Newer research data demonstrate that this no longer holds true universally.

METHODS

Five different studies were conducted involving a total of 562 female and 562 male students enrolled at the University of Hamburg. The volunteers selected as subjects in the experiments were predominantly in their early 20s, single, and with coital experience.

In the *first* study, black-and-white slides were used, portraying semi-nudes and nudes of the opposite sex, along with necking, petting, and coitus scenes (Sigusch *et al.*, 1970). In the *second* study, we utilized both black-and-white and color films as well as series of slides portraying necking, petting, and coitus (Schmidt and Sigusch, 1970). In the *third* study, we used two stories describing a sexual experience of a young couple; necking, petting, and coitus were described in detail (Schmidt *et al.*, 1973). In the *fourth* study, we used (as in the second study) black-and-white and color films and slide series, but the stimuli portrayed masturbation of a woman and masturbation of a man, respectively (unpublished). In the *fifth* study, we used four films of aggressive sexual content: one portrayed a sadomasochistic ritual, the second showed flagellant activity between two women, the third showed four men raping a woman in a bar, the fourth was a control stimulus showing nonaggressive sexuality (Ernst *et al.*, 1975).

In all five studies, each subject was left completely alone when he or she viewed the slides or films or read the stories. The experimenter was a member of the same sex as the subject. All data were accumulated by means of questionnaires which the subjects were required to fill out immediately subsequent to the experiment and 24 hr later.

GENERAL RESULTS

The general results of studies 1, 2, and 3 are published elsewhere in detail (Schmidt and Sigusch, 1973), and are only briefly summarized in this paper.

1. Women describe themselves as somewhat less sexually aroused after the pictorial and narrative stimulation than do men. However, these differences are only slight.
2. The great majority of women as well as men observe some sort of physiological-sexual reaction during the pictorial and narrative stimulation. These "objective" signs of sexual arousal are found in women about as often as in men.
3. Both women and men show a general and slight increase in sexual behavior (coitus, masturbation, orgasm) during the 24-hr period following exposure to pictorial and narrative stimuli as compared to the 24 hr before the exposure; there is also an increase in sexual fantasies and sexual desire. The extent of this activation varies only slightly between the sexes. Where significant differences between the sexes are noted, there is a general trend toward a greater activation among the women.
4. During the 24 hr after the pictorial and narrative stimulation, there is, in both men and women, a moderate tendency to incorporate the stimuli into their masturbation fantasies and a slight tendency to incorporate them into their fantasies during coitus. The influence of the stimuli on coital techniques (in the 24 hr after exposure) is in both women and men negligible.
5. Pictorial and narrative stimulation leads immediately to an emotional activation and agitation in both women and men; furthermore, it leads to an increase in emotional instability and to an increase in emotional tension; and, finally, it leads to emotional avoidance reactions. These avoidance reactions are stronger in women than in men.
6. Emotional activation and increase in emotional instability are still evident among both men and women 24 hr after their exposure to pictorial and narrative stimulation.

In sum, the pattern and intensity of reactions to explicit sexual stimuli are in general the same for men and women. When significant differences between the sexes are found, they represent merely minor shifts in the total pattern. These variations should not divert attention from the fact that women can react to the same extent and in the same direction as men.

This holds true at least for young, highly educated, sexually permissive subjects (university students). We do not rule out the possibility that in other social groups (older age levels, lower social strata) differences in arousability between the sexes still exist. However, we are concerned here with the fact that the effect of pictorial and narrative stimulation, *at least under certain social conditions,* is equally strong and quite similarly structured for both women and men.

To what extent does this generalization continue to hold true when the stimuli are varied? We have data on three stimulus variables: (1) affectionate vs. nonaffectionate stimuli, (2) aggressive vs. nonaggressive stimuli, (3) stimuli that allow projection vs. objectification (Money and Ehrhardt, 1972).

Affectionate vs. Nonaffectionate Stimuli

In study 3, we used two stories which differed in the degree to which affection was expressed. One story made it completely clear that the partners were solely interested in a sexual experience; the second story described the same sexual activities accompanied by affectionate desires.

It is generally assumed, and surveys of sexual behavior unanimously corroborate this (Ehrmann, 1959; Christensen, 1971; Sigusch and Schmidt, 1973), that the sexuality of women in Western societies is more dependent on affection than is that of men. This could lead to gender differences in sexual arousal in study 3. According to our findings, this not the case; the stories with and without affection do not have different effects on men and women (for details, see Schmidt *et al.,* 1973).

It is possible to argue that the stories which we used did not differ sufficiently with respect to "affection." Neither of the two stories was a "love story" or "romantic story." The two stories differed only in that in one of them more tenderness was expressed during sexual activity. Thus we cannot exclude the possibility of achieving greater differentiation in sex-specific responses using stories in which the emotional and social relationships of a couple are described in the same detail as sexual activities. However, our data do permit one important conclusion: *affection is not a necessary precondition for women to react sexually to sexual stimuli.* Even for stories which exclude and avoid any expressions of tenderness and affection (story 1), sexual arousal and sexual activation among females are as great as among males. This finding tends to refute the claim that women's sexual arousal is *basically* more dependent on affection than men's.

Aggressive Sexual Stimuli

In study 5, we investigated the reactions to films of aggressive-sexual content. Four films were used: (1) a sadomasochistic ritual, (2) flagellant activity

Table I. Aggressive Sexual Stimuli (Study 5): Ratings of Sexual Arousal, Favorable-Unfavorable Response, and Aggressiveness Immediately After the Experiment (Means, Standard Deviations)[a]

	Males			Females		
	Control (nonaggressive) ($N = 50$)	Sadomasochistic ritual ($N = 50$)	Rape ($N = 50$)	Control (nonaggressive) ($N = 50$)	Sadomasochistic ritual ($N = 50$)	Rape ($N = 50$)
Sexual arousal[b]						
M	5.8	3.7	5.4	5.2	3.2	4.2
s	1.7	1.9	2.1	2.1	2.0	2.7
Favorable-unfavorable response[c]						
M	3.4	7.2	7.1	3.9	7.9	7.9
s	1.7	1.7	2.0	1.5	1.6	1.2
Aggressiveness[d]						
M	2.1	2.6	4.7	1.9	3.4	4.9
s	1.6	1.9	2.7	1.7	2.3	2.7

[a]Significance of the sex differences according to Mann-Whitney U test: arousal: $p = 0.05$ for "rape," all others not significant; favorable-unfavorable: "control" not significant, "sadomasochistic" $p = 0.05$, "rape" $p = 0.05$; aggressiveness; all not significant. Significance of the differences between the films according to Kruskal-Wallis test: for all ratings and both sexes $p = 0.001$.
[b]Low value, low sexual arousal; high value, high sexual arousal.
[c]Low value, favorable; high value, unfavorable.
[d]Low value, low aggressiveness; high value, high aggressiveness.

Table II. Aggressive Sexual Stimuli (Study 5): Ratings of "Present Feelings" on a Semantic Differential Before and Immediately After the Experiment: Selected Items (Means)

	Control (nonaggressive) (N = 50)			Sadomasochistic ritual (N = 50)			Rape (N = 50)		
	Before	After	p^a	Before	After	p^a	Before	After	p^a
Males									
Composed–excited	2.6	4.5	0.001	2.5	3.9	0.001	3.0	4.7	0.001
Innerly agitated–innerly calm	4.7	3.4	0.01	5.1	4.0	0.001	4.4	3.3	0.01
Angered–amused	4.9	5.0	ns	5.1	4.3	0.001	4.9	3.9	0.01
High spirited–dejected	2.6	2.5	ns	2.4	3.3	0.001	2.6	3.9	0.001
Cheered up–depressed	3.3	3.0	ns	3.5	3.7	ns	3.4	3.2	ns
Unconcerned–shocked	3.4	3.3	ns	3.4	3.7	ns	3.2	4.1	0.001
Irritated–lazy	4.9	4.7	ns	5.4	4.7	0.01	5.2	4.3	0.01
Attracted–repelled	3.1	2.6	ns	2.6	4.8	0.001	2.9	4.8	0.001
Disgusted–pleased	4.6	5.4	0.001	4.9	3.4	0.001	4.6	3.3	0.001
Females									
Composed–excited	3.3	4.8	0.01	3.0	4.3	0.001	3.2	5.2	0.001
Innerly agitated–innerly calm	4.6	3.8	0.01	4.5	3.0	0.001	4.0	2.6	0.001
Angered–amused	5.2	4.8	ns	5.2	3.7	0.001	4.7	3.3	0.01
High spirited–dejected	2.4	2.5	ns	2.5	3.6	0.001	2.5	4.2	0.001
Cheered up–depressed	3.2	3.1	ns	3.3	3.6	ns	3.4	3.9	0.05
Unconcerned–shocked	3.5	3.8	ns	3.6	4.6	0.01	3.5	5.3	0.001
Irritated–lazy	5.1	4.7	0.01	5.0	3.9	0.001	5.2	3.4	0.001
Attracted–repelled	2.9	3.1	ns	2.9	5.6	0.001	3.0	5.8	0.001
Disgusted–pleased	4.8	5.1	ns	4.9	2.7	0.001	4.6	2.1	0.001

[a]Significance of the difference "before" and "after" according to sign test.

between two women, (3) group rape, (4) nonaggressive sex as control stimulus. (Film 2 is not further discussed here because it presented opposite-sex stimuli for the male subjects and same-sex stimuli for the female subjects.) Once again the male-female behavior patterns to the different stimuli were found to be quite similar (see Tables I and II).

1. Males and females respond to the control film with the strongest arousal (as compared to the other films) and with emotional activation, however with little emotional avoidance and with little aggressiveness.
2. Males and females respond to the sadomasochistic film with low arousal but with strong emotional avoidance, strong emotional labilization, dysphoric mood, and moderate aggressiveness.
3. Males and females respond to the rape film with relatively high arousal *and* strong emotional avoidance reactions, emotional labilization, dysphoric mood, and much aggressiveness. Thus this film evokes a strong conflict reaction — sexual arousal combined with strong aversion.

Significant sex-specific differences represent once again only minor shifts in this general pattern; women show even stronger emotional avoidance reactions than men to the aggressive films. Furthermore, the type of conflict experienced during the rape film seems to be somewhat different for men and women. (This finding did not emerge from our experiment but from group discussions with students about the rape film.) In women the rape film produces sexual arousal and, by identification with the female victim, fears of being helplessly overpowered. In men this conflict is more characterized by guilt feelings and dismay that they are stimulated by aggressive sexual activities incompatible with their conscious ideas of sexuality.

However, in our context the most important finding is that films which do not describe sexual aggression as a deviant or strange ritual (as in the sadomasochistic film) can induce strong sexual arousal in both men and women. Strong aggression in films of sexual content does not inhibit either men's or women's ability to react with sexual arousal. The aggression may even have an added sexually stimulating effect — once again for both men and women.

Projection vs. Objectification

Although women as well as men are able to respond to visual and narrative stimuli, Money and Ehrhardt (1972) describe differences in the imagery or experience of arousal. According to them, women's sexual arousal is programmed by projection, men's sexual arousal is programmed by objectification:

When he reacts to a sexy pin-up picture of a female, a man sees the figure as a sexual object. In imagery, he takes her out of the picture and has a sexual relationship .... The very same picture may be sexually appealing to a woman, but that would not mean that

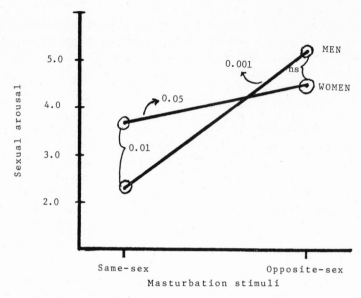

Fig. 1. Projection vs. objectification (study 4): Ratings of sexual arousal immediately after the showing of same-sex vs. opposite-sex masturbation stimuli
(means). Significance of mean differences according to t test. $N = 32$ in each cell.

she is a lesbian. Far from it. She is not in imagery bringing the figure toward herself as a
sexual object, as does the man. She is projecting herself into the picture and identifying
herself with the female to whom men respond. She herself becomes the sexual object.
What if the picture portrays a sexy male? The basic sex-specific difference still manifests
itself. Men are typically inattentive . . . they do not project themselves into the picture
and identify with the man there represented. Women, unable to identify with the female
figure, also do not respond to it as a sexual object. (Money and Ehrhardt, 1972, p. 252)

In study 4, we gathered data that can be related to this projection vs. objectification hypothesis. Films and slide series showing female masturbation or
male masturbation were used. The Money-Ehrhardt view leads to the following
hypotheses:

1. Women should report significantly *more* sexual arousal to same-sex masturbation stimuli than to opposite-sex masturbation.
2. Men should report significantly *more* sexual arousal to opposite-sex masturbation stimuli than to same-sex masturbation.
3. Women should report significantly *more* sexual arousal to female masturbation stimuli than men to male masturbation.
4. Men should report significantly *more* sexual arousal to female masturbation stimuli than women to male masturbation.

The average sexual arousal ratings (see Fig. 1) clearly confirm hypotheses 2
and 3. They are also in line with hypothesis 4, but the differences are not statis-

tically significant. However, hypothesis 1 — most important to the projection vs. objectification assumption — has to be rejected. The opposite is true: women (like men!) report significantly *less* sexual arousal to same-sex than to opposite-sex stimuli.

These data (especially with respect to hypothesis 3) are not to be explained by differences in projection-objectification tendencies; they probably result from greater inhibitions in adult men against same-sex stimulation and/or a greater "bisexual" capacity in women. Our data on favorable-unfavorable responses (see Fig. 2) are consistent with this inhibition hypothesis; the data show strong emotional rejection of same-sex masturbation stimuli by men, but not by women. It seems reasonable to assume that homosexual anxieties are stimulated

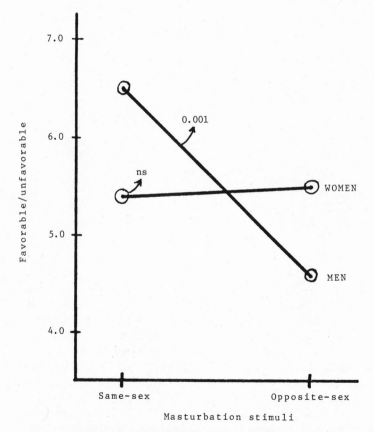

Fig. 2. Projection vs. objectification (study 4): Ratings of favorable-unfavorable response immediately after the showing of same-sex vs. opposite-sex masturbation stimuli (means). Significance of mean differences according to *t* test. *N* = 32 in each cell.

Table III. Projection vs. Objectification (Study 4): Sexual Reactions During the Showing of Same-Sex vs. Opposite-Sex Masturbation Stimuli and in the 24 hr Thereafter[a]

	Males (%)		Females (%)	
	Same sex (N = 32)	Opposite sex (N = 32)	Same sex (N = 32)	Opposite sex (N = 32)
Any sexual-physio-logical reaction	41	84	56	66
		0.001		ns
Masturbatory activi-ty during the expe-riment	3	22	19	19
		Too few cases		ns
Total orgasms in the 24 hr before and after the experiment				
After > before	16	38	22	25
After < before	25	19	9	9
Equal	59	44	69	66
		0.10[b]		ns[b]
Wish for sexual acti-vity in the 24 hr before and after the experiment				
After > before	13	44	38	50
After < before	34	19	9	16
Equal	53	38	53	34
		0.01[b]		ns[b]

[a]Significance of the differences between same-sex and opposite-sex stimuli according to the χ^2 test.
[b]Categories "after < before" and "equal" were combined for calculating χ^2 values.

more quickly in men than in women; both sexes probably respond sexually to same-sex stimuli, but this reaction is more anxiety laden and therefore more easily suppressed in the male.

The greater inhibition induced by same-sex stimuli is also to be seen with regard to sexual reactions during our experiment and sexual activation in the 24 hr after the experiment (see Table III). Women showed about equally strong re-actions to same-sex and opposite-sex stimuli, whereas men responded with de-finite lower arousal to same-sex than to opposite-sex stimuli.

Thus reactions to same-sex stimuli, unlike reactions to most other stimuli studied to date, are gender dimorphic. The explanation for this is still unknown.

SOME CONCLUSIONS

Our data show more similarities than differences in male-female sexual arousal by pictorial and narrative stimuli. These similarities have become more visible during the decades since the Kinsey studies. They are thus one symptom of the abovementioned convergence of male and female sexual behavior patterns. I want to comment on the sociological factors responsible for this convergence.

Christensen (1966, 1971) has demonstrated the parallelism between sexual liberalization and the convergence of male and female sexuality. In a heterosexual-oriented society there is no far-reaching liberalization of sex possible without giving up the double standard and culturally conditioned sexual inequality because male and female sexual behavior are reciprocally dependent. Thus, sociologically speaking, the sexual liberalization process we have witnessed during the last years (Schmidt and Sigusch, 1972; Hunt, 1974) is one powerful factor in the convergence, and we have to explore the reasons for liberalization to explain the changes in male and female sexuality.

The liberalization of sexuality can be observed in many, perhaps all, affluent late industrial societies. It is *possible* in these societies because the traditional sexual repression is no longer functional, and it may even become *necessary* as the traditional repression becomes more and more dysfunctional.

Sexual liberalization is *possible* because an industrial society needs to follow an ethic of renunciation only during the phase of capital accumulation, during which all produced profits have to be reinvested and the consumption of the produced goods has to be postponed (Israel *et al.,* 1970). But an ethic of renunciation will no longer be functional and will therefore become renounceable in an industrial society with a high standard of living where an increase of output is achieved by better-developed technology, where work becomes much easier physically, and where working hours are reduced. Such a society may lessen the control over sexuality, for it has less interest in control.

Sexual liberalization may even be *necessary* in late industrial societies as the traditional limits on sexuality become dysfunctional in various ways. Sexuality and the social activities linked up with it become more and more necessary to fill the steadily increasing spare time; sexuality is used to open new markets and thereby becomes useful for increasing consumption and profits. So, for example, the liberalization of adolescent sexuality is accompanied by the development of a special juvenile market with an extremely sexualized advertising. The liberalization of the antihomosexual laws in West Germany opened the market to the homosexual subculture. The reduction of prejudices against homosexuals has probably prepared men for the higher consumption of a more female fashion. A liberalization of the traditional ways of living together sexually becomes necessary as the traditional patriarchal family with its traditional division of labor no longer suits the importance of women in the labor force. Furthermore, societies that are dependent on fast and steady consumption (like late industrial societies) and that therefore teach behavior patterns directed at immediate and frequent satisfaction of real or artificially produced needs cannot prevent a transfer of these behavior patterns to the field of sexuality. Sexual norms become more and more influenced by consumption values as opposed to production values (Gagnon and Simon, 1973). Thus sexual liberalization does not endanger the social structure of late industrial societies; sexual liberalization is instead an integral part of these societies — it is affirmative and thus really conservative.

The traditional gender roles are generally changing. More women get a higher education, more participate in the labor force, more want to have (and do have) a professional career, more reject the goal of being merely a housewife and mother. Women start to emancipate themselves from economic and socioemotional dependency on husband and family. The erosion of gender roles has just begun and is especially marked in the middle class. The importance of these changes for sex differences has not been investigated empirically until now. However, one can assume that the increase in role flexibility has clearly influenced sexual gender roles.

Social equality and thus sexual equality for women are functions of their participation in public life and public production. This participation is slowly growing and this, too, is surely a powerful factor in the convergence of male-female sexuality.

Group Discussion

Dr. Rose noted that the Schmidt studies must rely on self-reports by subjects and asked whether the desire of respondents to give appropriate responses might not contaminate the data. Dr. Schmidt replied that it might be better actually to measure physiological responses in some kinds of studies, but that when comparing male with female responses some ambiguity may arise in direct-measurement studies. Those measures which can be used in both males and females do not measure specifically sexual response but only general physiological arousal. Those which measure specific sexual response — for example, penile plethysmography — can be applied only to one sex. Dr. Rose noted that even if used in only one sex direct measurement could confirm or impugn the accuracy of self-reports. Dr. Lewis pointed out that directly measured responses could also be contaminated by a subject's desire to respond appropriately. Dr. Schmidt called attention to the concrete nature of the self-reports in his study — for example, answers to the question "Have you had an orgasm in the last 24 hours?" The responses to such questions may be more reliable than the responses to general questions about sexual arousal.

Dr. Cole noted that the Schmidt findings were consonant with what appeared to be happening among the 6000 men and women exposed to explicit sexual materials during University of Minnesota Medical School programs.

A detailed discussion of methodology followed. It was pointed out, for example, that sexual response can be measured directly — in the male through penile plethysmography and more recently also in the female through vaginal plethysmography. Less intrusive indirect indices of sexual arousal can also be used. Dr. Schmidt concluded the methodological discussion by noting that a highly sophisticated set of research techniques might indeed modify his findings in certain respects — but that this is *not* the most pressing task of sex research in its present state. Vast areas remain wholly unexplored. There is need to fit various portions of the puzzle together. It is such tasks as these rather than the repetition of the same work in more and more sophisticated ways that should be the focus of attention.

Dr. Schmidt's finding that women in his sample were more readily aroused than men by same-sex stimuli was contrasted by one participant with data indicating that homosexual *activity* is engaged in by more men than women. Dr. Lipman-Blumen pointed out in this connection that the antihomosexual taboo is much more stressed in the upbringing of boys than in the upbringing of girls; hence boys have a greater stake in avoiding same-sex arousal. Dr. Green agreed; a same-sex encounter strikes more at the "core sexual identity" of the male than of the female in our culture. A female can remain quite feminine despite a same-sex experience; the experience is therefore less devastating to her. The word "tomboy" has a much less negative connotation than the word "sissy." It was also pointed out, however, that some evidence conflicts with this model — such as the common experience of group masturbation among young boys. Mr. Brecher suggested that this may be related to different ways of defining homosexual activity. Thus in some American subcultures only the male who plays the passive or "insertee" role in male-male contacts is labeled homosexual; the male who plays the active or "inserter" role can and does continue to consider himself heterosexual.

Dr. James questioned the relevance of much of this discussion to the human condition. Suppose that by means of highly sophisticated and unchallengeable methodology it was possible to establish certain specific differences between male and female sexuality. What then? Of what use would the information be? In the vast field of sex research, those projects most merit support which give promise of leading to human betterment.

A generation earlier, it was pointed out, the social utility of these studies would be abundantly clear; their findings could be used to prevent homosexuality, transsexuality, and a long list of other modes of behavior which people wanted to prevent. This motivation is now beginning to fade; and the exciting news presented in the Schmidt study is that differences between male and female sexuality are simultaneously beginning to fade — at least in sexually experienced West German college students.

Heterosexual Dysfunction: Evaluation of Treatment Procedures[1]

Diane S. Fordney-Settlage, M.D., M.S.[2]

INTRODUCTION

As professional and public awareness of the potential for treatment of sexual problems increases, there are an ever-expanding number of available therapeutic services. As researchers as well as clinicians, we find ourselves in the dilemma of maintaining services without having good research data to support our treatment techniques and therapy. So-called successful sexual dysfunction techniques are often based on isolated case reports. However, with few exceptions these techniques and reports of techniques generally lack retrospective and prospective predictive ability, standardization, objectivity parameters, detailed outcome evaluation, and adequate follow-up. Often they tend to be anecdotal and limited in their ability to provide usable information to professionals in the same or allied areas.

The general categories of reported treatment approaches can be labeled as indirect, direct, and combined or eclectic. More specifically, indirect methods are primarily insight-oriented and supportive techniques. They may be used for individuals in psychotherapeutic or psychoanalytic modalities and for units in marital or conjoint counseling. The direct category is those techniques which deal specifically with sexual concerns or problems and have primarily been utilized with individuals. These consist of behavior modification techniques utilizing models developed in the treatment of other symptomatology.

The third category is that of combined (or eclectic) approach which currently is utilized almost exclusively for male-female units and consists of several

[1] This paper was presented at the conference, "Sex Research: Future Directions," held at the State University of New York at Stony Brook, Stony Brook, New York, June 5-9, 1974.
[2] Department of Obstetrics and Gynecology, School of Medicine, University of Southern California, Los Angeles, California 90033.

basic components. These components are as follows: (1) a sexually conjoint male-female therapy team works with a single male-female problem unit; (2) the therapists assume the responsibility for the sexual activity of the dysfunctional unit to alleviate anxiety; (3) the therapy team provides sexual information and permissive sexual attitude modification; (4) there is disruption of defective sexual behavior patterns by replacement with more effective patterns utilizing desensitization and progressive sexual conditioning techniques; (5) there is development of intraunit sexual communication based on replacement of defective communication patterns with more effective methods provided in the therapy setting; (6) such teams also provide nonsexual marital counseling, or relationship counseling, and sometimes individual counseling where such problems are obstructive to the goals of sex therapy.

There is today very little information as to why various therapy approaches are effective with specific cases. We have not yet isolated the essential vs. the sufficient variables necessary to promote sexual functioning. Nor have we isolated those techniques which produce maximal effect alone or in combination, minimal cost in terms of therapist time and patient expense, or ways in which maximal efficiency can be produced. Some of the major problems facing people who provide sex counseling involve criteria by which appropriate screening or selection of patients for sex counseling can be made. Are there differences in etiology and suitability for treatment by presenting problem, or are these differences related to other characteristics in the patient populations? Similarly, what are the bases for the selection of appropriate therapy?

What is needed is a research design whereby varying treatment approaches and the influence of the wide variety of variables on the complaint of sexual dysfunction can be evaluated. An idealized treatment design requires several areas of attention. The first of this is *pretreatment assessment* of the study population as dysfunctional units when the dysfunction exists in a long-term or monogamous relationship. Pretreatment assessment would need to include clinical evaluation related to symptom specificity, goals and expectations of the patient, affective responses of the individual, affective responses within the unit, and a medical history and examination in respect of sexual function, with correlations to objective pretreatment data.

A second major factor would include *consistent objective data collection.* Data would include such items as demographics of the study population, sexual behavior inventories, and etiological factor assessment through the use of surveys and psychometrics which quantify sexual information, sexual attitudes, sexual interpersonal reactions, nonsexual interpersonal reactions, and intrapsychic factors. Unit assessments would also be included in this group as to expectations for satisfaction of the unit, reaction patterns, and integrity of the unit.

A third facet of this design would include *intratherapy evaluation,* both clinical and utilizing standardized objective scales. Finally, there must be *post-*

therapy evaluation in terms of both *initial* and *long-term outcome.* Considerations of this post-therapy evaluation would include the presence or the absence of the initial symptoms for the individual and for the unit. It would include whether or not the stated goal was achieved for the patient or for the unit in question. This evaluation should also include overall sexual function in terms of improvement or lack of improvement whether or not the symptoms were alleviated, and this overall sexual satisfaction again must be measured in terms of the individual and the unit. Finally, the overall unit relationship must be evaluated. Post-therapy evaluation would include initial and long-term comparative objective test data. Data collected in conjunction with standardized treatment approaches and combination of approaches can then be utilized to isolate their necessary or sufficient variables as they relate to symptoms and to individuals.

Consistent with the goals of maximizing therapeutic effectiveness, there are identified treatment procedures which are controllable and which have been found useful in actual practice. These include the forms of administering therapy — to individuals alone with same- or opposite-sex therapists, or in groups of individuals with same- or opposite-sex therapists or conjoint therapy teams. Results can be compared to treatment of units by a conjoint therapy team alone, or in groups, or by a single therapist of either sex, or by dividing the units and treating their members separately. A combination of the effects of partially conjoint and partially individual therapy can also be evaluated. Immediately assessable treatment techniques isolated for an individual (or for a unit, by not utilizing ancillary approaches concurrently) include such variables as general sexual infomation and attitude discussion, sexual communication and assertion training, sexual techniques training, self-assertion and self-esteem training (nonsexual in focus), identification and disruption of destructive behavior which are primarily nonsexual in focus, and pharmacological, medical, or surgical treatment.

PILOT STUDY

The basic design for an approach to evaluation of heterosexual dysfunction and treatment outcome has evolved, over the past 2 years, from our treatment experience with a lower-socioeconomic-status patient population. Study of this group constituted a pilot program at our institution to evaluate sexual dysfunction treatment in this population and identify some of the major variables associated with problems they presented. Motivation is one such major variable. It must be understood, however, that motivation in this population includes such factors as partner resistance to therapy, the problems of interference of therapy with maintenance of job security, practical factors such as child care and transportation, as well as the multiple, nonspecific motivational factors. Table I indicates the numbers of patients lost during the therapeutic process at

Table I. Proportion of Sample Requesting But Failing to Complete Sex Therapy

			Losses		
	N	From contact to appointment	From not keeping appointment	From screening out	From premature closure
Contact	339				
Given initial appointment	292	47 (14%)			
Kept initial appointment	212		80 (27%)		
Entered treatment	197			15 (7%)	
Finished treatment	147				50 (25%)

Table II. Identifiable Reasons for Failure to Complete Sex Therapy by Level of Involvement

1 Between contact and initial appointment:
      38 — could not reach
      7 — separated from partner, no need (4) or desire (3) for therapy
      2 — not separated from partner (stated they had no current problem)
2 Failed to keep initial appointment:
      23 — husband forbade
      21 — too much trouble
      16 — failed 3 consecutive appointments
      16 — could not reach
      4 — hostile (needed rescheduled appointments)

3 Screened out:
      6 — to marriage counselor
      5 — no sexual problem
      2 — psychotic
      2 — to psychiatric individual service

4 Lost after 3-4 visits:
      19 units
      8 individuals } separated from partner

      11 units — male partner resistance

      4 units
      1 female individual } therapist antipathy

      5 units — to marriage counselor
      2 units — left area

Table III. Major Recurrent Identifiable Factors Associated with Failure to Complete Sex Therapy at All Levels of Involvement

Evidence of severe, disruptive relationship distress:
45 (14% of total)

Evidence of male partner resistance:
34 (10% of total)

our institution. Table II lists the reasons for those losses, and Table III lists, from the group of patients lost to treatment, major areas of difficulty.

The characteristics of the treated patient population are shown in Tables IV and V. These are based on presenting complaint and do not include coexisting problems or problems identified once therapy began. Because the patient sample was generated largely from a gynecological clinic and the Sexual Problems Clinic is based in a gynecological hospital, the proportion of females is high. There was a high percentage of women whose initial complaint was dyspareunia, particularly among the Chicana subgroup. This may represent the severity of sexual dysfunction before consultation was requested in this particular population.

Table VI reveals the frequent incidence of coexistent male problems in units treated for a primarily female-identified problem.

There was, however, a group of women who were treated initially without their partners, because of partner reluctance in most cases. Those whose partner entered treatment at some point after it had been begun with the female individual were then considered to be unit treatment patients. For one-third of those patients who began therapy without partners, sufficient progress was made to enable the unit to enter therapy, adding strength to the hypothesis that single partner treatment may be worthwhile. Of those 52 treated alone throughout, all

Table IV. Characteristics of the Population Treated with Female-Identified Complaints ($N = 175$)

	Vaginismus	Dyspareunia	Orgasmic disorder	All
Age				
Median	22	29	27	28
Range	18-28	18-58	23-54	18-58
Ethnic group				
White	3	21	25 (60%)	49
Black	8	26	13	47
Chicana	18 (50%)	49 (51%)	2	69
Oriental	–	1	2	3
Arabic	7 (20%)	–	–	7
Annual income				
< $5000	12	73	15	100
< $10,000	24	24	27	75

Table V. Characteristics of the Population Treated with Male-Identified
Complaints (N = 22)

	Secondary inpotence	Premature ejaculation	Ejaculatory incompetence	All
Age				
Median	43	28	26	36
Range	26-68	24-36	25-29	24-68
Ethnic group				
White	8	2	1	11
Black	6	1	1	8
Chicano	0	0	3	3
Annual income				
< $5000	5	0	3	8
< $10,000	9	3	2	14

women complaining of either vaginismus or dyspareunia had alleviation of that symptom with individual treatment except for premature closure of treatment for two patients with primary coital anorgasmia and four patients with secondary coital anorgasmia. However, 24 of the 35 patients with vaginismus and dyspareunia also had an anorgasmic disorder. Of the ten with absolute anorgasmia, seven became coitally orgasmic. Of the five with primary coital anorgasmia, four became coitally orgasmic. Of the nine with secondary coital anorgasmia, only two became orgasmic. Clearly, for secondarily anorgasmic women maintaining the sexual unit in which the symptom developed, sex therapy only of the woman was relatively ineffective.

Table VII summarizes a therapist-evaluated classification of patients with identified female problems and illustrates the coexistence of multiple sexual dysfunction problems. In those patients with vaginismus, all were primarily coitally anorgasmic and all had dyspareunia irrespective of whether the vaginismus was absolute, such that penetration was impossible, or relative, so that it was possible only with great difficulty. Of the sample of women presenting with dyspareunia, 59% had a prior orgasmic disorder.

Of the patients presenting with anorgasmia but without history of dyspareunia, only 12% had developed secondary orgasmic disorders. This again indicates that dyspareunia tends to be associated with prior lack of sexual satis-

Table VI. Incidence of Coexistent Male Dysfunctions in
Female-Identified Dysfunctions

Female-identified problems, unit treatment =	97
Coexistent male problems	= 38 (39%)
Total problems	= 135

Table VII. Summary of Female-Identified Dysfunctions Treated

Female problems	N	Percent of total
Vaginismus	36	21%
Absolute	22	
Relative	14	
Orgasmic disorders	42	24%
Absolute	19	
Coital	23	
Primary	18	
Secondary	5	
Dyspareunia	97	55%
With absolute		
orgasmic disorder	19	
With primary coital		
orgasmic disorder	16	
With secondary coital		
orgasmic disorder	22	
(Unceasing libido	1)	

faction or difficulty. In the absence of organic disease, this presenting complaint may well represent an increased severity of sexual symptomatology and disorder. Of those women who had dyspareunia alone, fully one-third were found to have either etiological or contributing organic problems such as subacute vaginitis, prior inflammation of pelvic organs with fixation of the ovary, or postsurgical and postepisiotomy difficulties. All of the women who had organic factors contributing to dyspareunia had the secondary type and most of them remained coitally orgasmic.

Of those patients who had dyspareunia with a secondary coital orgasmic disorder, approximately one-third had some organic contributing factor. Treatment of this particular subgroup of patients was successful with a combination of medical and surgical intervention, and use of the sexual dysfunction training therapeutic model. Full resumption of coital orgasm resulted.

Table VIII shows the interaction between females' disorders and their partners' disorders. It can be seen that continued treatment was generally successful in removing the presenting complaint. However, of more interest are those cases listed under "premature closure," where therapy was terminated before completion. No premature closure occurred beyond the fifth session, less than half way through a normal course of treatment. In some instances, the premature closures occurred as early as the second session. When the complaint was of orgasmic disorder or dyspareunia, premature ejaculation in the male (P.E.) was the *only* male dysfunction associated with primary coital anorgasmia in the female, and

Table VIII. Results of Female-Identified Dysfunctions Treated by Unit
Therapy

	n	Premature closure	Failures
Female orgasmic disorders			
Absolute	12	0	–
+ P.E.	3	0	–
Primary coital	4	1	1
+ P.E.	4	3	
Secondary coital	8	6	–
+ P.E.	4	4	–
+ Secondary impotence	3	1	
Dyspareunia	12	–	–
+ Absolute anorgasmia	5	–	–
+ P.E.	3	1	–
+ Primary coital anorgasmia	7	1	–
+ P.E.	4	2	–
+ Secondary coital anorgasmia	15	7	–
+ P.E.	9	8	1
+ Secondary impotence	2	1	–
+ Ejaculatory incompetence	1	–	–
Vaginismus	13	–	–
+ P.E.	9	1	–
+ Secondary impotence	12	–	–

was present in one-half of cases of dyspareunia and primary coital orgasmic disorder alone. The premature closures in this particular group were predominantly in those cases in which premature ejaculation was also a factor. With dyspareunia plus secondary coital anorgasmia or pure secondary coital anorgasmia, premature ejaculation, secondary impotence, and ejaculatory incompetence were all noted. Of these, premature ejaculation was the most common male dysfunction. In women with secondary coital anorgasmia, with or without dyspareunia, premature ejaculation was almost invariably associated with premature closure. This was not the case with secondary impotence or ejaculatory incompetence. It would appear that that particular combination is associated with failure to follow through a therapeutic program. It was repeatedly noted by therapists in these situations that the amount of marital disharmony was extremely high, as were frustration with sexual retraining and intraunit hostility.

In contrast, treatment units with a primary complaint of vaginismus showed evidence of premature ejaculation and secondary impotence. The premature closure phenomenon was not seen in this group of patients and the duration of the problem was, on average, much less than with the other groups. Additionally, the majority of these patients had initially sought infertility evaluation and had a source of motivation not found in other groups.

Table IX. Male-Identified Dysfunctions Treated by Unit Therapy

Male-identified problems: unit treatment = 15
Coexisting female problems = 4 (27%)

Total problems: 19	n	Premature closure	Failures
Secondary impotence	6	–	1
+ dyspareunia	1	–	1
+ coital anorgasmia	1	–	–
Premature ejaculation	–	–	–
+ coital anorgasmia	1	–	–
+ dyspareunia	1	–	1
Ejaculatory incompetence	5	–	1

In treatment of female-identified problems utilizing unit treatment, although the majority of the units were treated by a male-female cotherapy team, approximately one-third were treated by either a female or a male therapist alone. The premature closure rate did not apparently differ in this pilot sample between therapist models and the failure rate did not differ appreciably. Significance cannot be drawn from this preliminary study because of the small sample size in the various subgroupings and the individualization of the treatment modalities utilized.

Our experience with male-identified problems has been slight, as shown in Tables IX and X. In the case of secondary impotence, the existence of female problems preceded the development of the male problem. With premature ejaculation the female problems were concurrent with the unit formation. Two striking features in the treatment of male-identified problems form the basis for including these tables despite the paucity of data. The first is that there was no premature closure either in unit treatment or in individual treatment if it was a male-identified problem. The second, probably related feature was that of seven male individuals treated without partners none had a regular partner to form a unit for treatment.

These features differ markedly from the data accrued with female-identified problems. There may be greater passivity on the part of women to involve their partners when they identify themselves as the dysfunctional partner. Or

Table X. Male-Identified Dysfunctions Treated Without Partners

	n	Failure	Treatment
Premature ejaculation	1	–	Male-female team
Secondary impotence	6	2 (organic)	Male-female team

there may be greater willingness for the female partners of dysfunctional males to enter into the therapeutic process than *vice versa*. No other data from this clinic can be presented because patients were seen primarily in a clinical service and training setting rather than in a research format. However, some recurring problems were observed. Dysfunctional women with coital anorgasmia showed lack of sexual as well as reproductive self-knowledge, and engaged in very little experimentation or coital variety. They often felt self-revulsion with respect to sex organs. They had difficulties in self-assertion, predominantly around sexual issues or expression of feelings generated by sexual failure. Their body image and self-image were generally poor. For presenting male individuals, anxiety and feelings of total responsibility for the dysfunction were paramount. Among units treated, sexual stereotyping or "machismo" was much less frequently encountered than was general inability to communicate sexual needs and feelings.

Where premature ejaculation existed coincident with a female-identified problem, it was not usually considered as a problem to that unit, although it was treated tangentially in the retraining process. Where the female-identified problem was secondary, essentially all units revealed serious relationship difficulties. Those in which the secondary symptom also occurred secondarily in the unit were associated with the highest premature closure and failure rate. Those who continued in therapy generally were able to remove the sexual symptom and resolve the relationship problem.

It appears that in this patient group the problems of self-image, body image, and sex role stereotyping are higher than in middle-class populations. However, specifically sexual items or relationship difficulties show more similarities to other reported samples than dissimilarities.

Data and impressions gained during the study of this sample indicate the need for greater specificity of sample characterization and better clinical evaluations. The many unexpected outcomes further emphasize the need for better understanding of the therapeutic process. For example, sexual problems are sometimes resolved either prior to detailed evaluation or following educational sessions or prior to therapeutic counseling; problems sometimes recur 6 months after completion of sexual therapy, or are eliminated 3-6 months following a treatment failure.

CLINICAL EVALUATION

Usually patients are classified according to their presenting complaint. The patient's complaint does constitute a presenting complaint but is rarely adequate for either diagnostic or prognostic evaluation. Part of the difficulty has to do with a lack of symptom specificity. Broad symptom classifications exist but confusion persists because of a great range of variables which affect a symptom, symptoms not classified, multiple symptom presentations, and the introduction of new terminology which has only partial professional acceptance. Patients often come in with concern about a sexual problem, which may cause difficulty to the unit, without identifying other preexisting difficulties to which that sexual problem is a secondary reaction. Categorization of problems by prior data would make possible screening for therapeutic modalities in conjunction with or instead of sex therapy, immediate recognition of possible organic variables immediately alterable, individual or unit destructive situations associated with

Table XI. Diagrammatic Model for Heterosexual Dysfunction Symptom(s): Overview

1. Individual, female or male

  a. Primary   Secondary
     symptom   symptom

     1. Has essentially   1. Has developed
        always existed for   after a life
        the individual   period free of
                                             that condition

  b. Symptom group

     1. Based on stage(s) of the sexual response cycle
        where symptom(s) appear

  c. Activity group

     1. Refers to sexual or physical stimulation

  d. Situational and partner dependence

     1. Refers only to secondary symptoms

2. Female and male unit:
  a. Primary symptom(s)   or   secondary symptom(s)

     1. Has essentially   1. Has developed
        always existed   after a life period
        within the unit   free of that
                                            condition in the unit

  b. Symptom group(s)
     1. Temporal relationships

  c. Activity groups
     1. Presence or absence within the unit
     2. Initiation patterns
     3. Reactions of unit to the activity
     4. Differences in desirability of unit activity
     5. Reasons for absent activity

  d. Situational dependence
     1. Relationship to unit nonsexual distress
     2. Relationship to individual distress of either partner
     3. Relationship to unit-affecting external variables
     4. Conditions of greater sexual satisfaction for unit

Table XII. Detailed Symptoms of Female Individuals

Type	Symptom group	Activity group	(Secondary symptoms only) Situational or partner dependence
Primary	1. Excitation disorders a. No desire b. Deficient desire c. No sensation d. Deficient sensation e. Painful sensation f. Lubrication deficiency	1. Solitary a. Nocturnal b. Dream state c. Fantasy d. Nonsexual physical contact e. Nongenital stimulation f. Genital stimulation	1. Situational effects a. Relative privacy b. Physical well being c. Emotional well being d. Vacation e. Stress
	2. Intromission disorders a. Absolute vaginismus (no penetration) b. Relative vaginismus (painful forced penetration) c. Introital pain d. Excitation loss e. Anxiety provocation	2. With partner a. Nocturnal b. Dream state c. Fantasy d. Nonsexual physical contact e. Nongenital stimulation f. Genital stimulation Oral Manual Mechanical Coital Anal	2. Partners a. Regular b. Other feeling relationship c. Casual d. Same sex e. Other
Secondary	3. Thrusting disorders a. Introital pain b. Vaginal barrel pain c. Point-specific pain Introital Deep thrust d. Generalized deep-thrust pain e. Anxiety provocation		
	4. Orgasmic disorders a. Uncertain b. Conditional c. Infrequent d. No orgasm, sham orgasm		
	5. Resolution disorders a. Pain (location) b. Anxiety c. Hostility d. Depression		

treatment failures, increased prognostic and predictive capacity for each pre-sented problem, and logical selection of sexual treatment approaches. The dia-grammatic model of symptom(s) classification presented in Tables XI, XII, XIII, and XIV provides one approach.

Once a temporal arrangement of symptoms is made this way, a weighted evaluation of factors presented by clinical assessment, utilizing strictly the case history, can be made. The major factors to assess, for the female, the male, and the unit, would include the following:

1. Sexual information and attitude deficiency.
2. Sexual experience deficiency.
3. Sexual communication deficiency.
4. Negative experience.
5. Regressive sexual behavior.
6. Destructive reaction patterns.

Table XIII. Detailed Symptoms of Male Individuals

Type	Symptom group	Activity group	(Secondary symptoms only) Situation or partner dependence
	1. Excitation disorders a. No desire b. Deficient desire c. Erectile inability d. Erectile deficiency e. Erectile pain	1. Solitary a. Nocturnal b. Dream state c. Fantasy d. Nonsexual physical stimulation e. Nonsexual genital contact f. Genital stimulation	1. Situational effects a. Relative privacy b. Physical well being c. Emotional well being d. Vacation e. Stress
Primary	2. Intromission disorders a. Loss of erectile ability b. Anxiety provocation	2. With partner a. Nocturnal b. Dream state c. Fantasy d. Nonsexual physical contact c. Nongenital stimulation f. Genital stimulation Oral Manual Mechanical Coital Anal	2. Partners a. Regular b. Other feeling relationships c. Casual d. Prostitutes e. Same sex f. Other
	3. Thrusting disorders a. Loss of erectile ability b. Pain c. Anxiety provocation		

Secondary	4. Orgasmic disorders
	a. Ejaculatory incompetence
	b. Conditional ejaculatory incompetence
	  Idiosyncratic requirement
	  Additive techniques
	c. Preerectile ejaculation
	d. Excitatory phase ejaculation
	e. Intromission ejaculation
	f. Premature coital thrusting ejaculation		

	5. Resolution disorders
	a. Pain
	  Penile
	  Pelvic
	b. Anxiety
	c. Hostility
	d. Depression		

Table XIV. Detailed Symptoms of Unit

Type	Symptom group	Activity group	Situation dependence
Primary	1. Excitation disorders Female Male	1. Nocturnal dream state arousal Female Male	1. Identified importance to unit of decreased sexual satisfaction a. External variables Privacy Children Relatives Income Job Lack of time b. Nonsexual unit distress Expressed anger Tears Withdrawal Avoidance Fear Sharing lack Withholding bargaining Infidelity
Secondary	2. Intromission disorders Female Male 3. Thrusting disorders Female Male	2. Fantasy Female Male 3. Nonsexual physical contact Female Male Desired Adequate Initiated Reaction If absent, why	2. Conditions of increased satisfaction (specify) Female Male Female and male 3. Unit observed interaction (yes-no or ± format)
	4. Orgasmic disorders Female Male	4. Nongenital stimulation Female Male Desired Adequate Initiated Reaction If absent, why	
	5. Resolution disorders Female Male (Label primary symptoms P_1, P_2, etc., secondary symptoms S_1, S_2, etc., for temporal assessment.)	5. Genital stimulation type Female Male Oral Manual Mechanical Desired Adequate Initiated Reaction If absent, why	
		6. Coital contact Female Male Desired Adequate Initiated Reaction If absent, why	

   7. Nonsexual interpersonal distress.
   8. Intrapsychic factors.
   9. Inadequate sexual self-concept.
   10. Damaged self-concept.

   This weighted evaluation could be made on a 4-point scale of the magnitude of the problem for the individuals and for the unit, thereby focusing on sex therapy appropriate vs. sex therapy inappropriate predictions, major areas of difficulty for the unit, and the areas which require individual therapy. Utilization of such a historical outline has the advantage of placing symptoms under the phases of the sexual response cycle (excitation, plateau, orgasm, and resolution) and helps to identify the initiating problems as well as the reaction patterns that develop in response to the problem. Organization of history in terms of activities both alone and with partner enables us to assess the amount of experience, exposure to sexual experimentation, and mutual confidence, allowing evaluation of inhibition, ignorance, assertion, and sexual attitude. The history of secondary symptoms showing situational or partner dependence usefully describes the organization of daily life and relationship to partner, i.e., effectiveness of the *unit* in sexual function. The goal here is to identify differences between an individual's sexual function and dynamic or organic situations which have no bearing to a particular partner as opposed to difficulties encountered by the unit which are nonsexual in nature.

   An overview of the general discriminatory usefulness of the classification can be divided into primary and secondary symptoms for the individual and the unit. First, individual primary symptoms are also unit primary symptoms. These symptoms antedate the unit formation and reflect an individual deficiency which is unrelated to the partner. These include deficient sexual information, erroneous sexual attitudes, and lack of or negative sexual experience. Additionally, a deficient sexual self-concept and possible intrapsychic factors for the individual may be involved. This situation stresses the unit and may impair unit bonding, which bonding has been insufficient to overcome the problem. Prognostically, such a situation suggests a need for additive, individual-specific techniques which do not require unit involvement. Potentially, improvement of the individual may significantly improve the overall unit relationship.

   Second, unit primary symptoms, with individual secondary symptoms, appear to place more emphasis on unit factors than on specific individual factors. These indicate sexual deficiency even under optimal or early unit conditions, where the sexual symptom is probably a major stress to the unit. Sexual information, attitudes, and experience of both individuals may operate to produce the symptom. An alternative is that unit formation may be faulty, or unit nonsexual satisfaction may be low. Relief of the unit primary symptoms may improve the overall unit relationship by removing one stress. Learning communication skills and coping techniques can improve other relationship stresses. Therapy for both individuals in a unit is indicated here.

Third, individual secondary symptoms with unit secondary symptoms indicate difficulties in the unit bonding and reaction patterns. If a unit primary symptom exists, secondary symptoms reflect destructive sexual reaction patterns which damage the overall relationship. Therapy approaches must emphasize disruption of these patterns and initial alleviation of the secondary symptom. An individual or unit primary symptom may exist which is not perceived as a problem by the individual or unit. For example, intromission and coital thrusting pain may exist in a primarily nonorgasmic female who had no expectation of orgasm. Here, the patient's perceptions, as they influence therapy evaluation and goals, must be considered. If, however, unit secondary problems are also individual secondary problems outside the unit activity, severity of individual damage or individual specific factors may be present. Additional individual techniques merit evaluation in this instance. If the unit secondary problems are limited to the unit, the unit integrity and nonsexual relationship patterns must be thoroughly assessed for appropriateness of sex therapy as opposed to conjoint or marital therapy. If appropriate for sex therapy, it is essential to treat the unit.

In summary, primary problems, whether individual or unit, imply deficient sexual information, restrictive sexual attitudes, deficient or negative sexual experience, inadequate sexual communication, individual self-concept deficiencies, and possibly individual intrapsychic factors. Secondary problems, whether individual or unit, imply a negative experience, regressive sexual communication and/or behavior, nonsexual interpersonal distress, secondarily damaged self-concept, and destructive reaction patterns. This clinical evaluation can be made from an individual interview of both the male and female of the unit and a unit interview in which unit factors are emphasized. Next would come a total physical examination of each individual with laboratory assessments for routine medical screening and data-based other evaluation. The initial clinical observation and categorization would then be completed.

OBJECTIVE DATA COLLECTION

Even when collecting clinical information in a consistent fashion, it is important to develop measures to identify factors which may affect patient perception, symptom etiology, and treatment focus. Very little objective information exists relating patient dysfunction and perception to etiological variables, to differences in patient groups, or to different treatment approaches or outcomes. The optimal time for collecting baseline data is immediately following the initial contact, which should be limited to a brief patient unit description of the problem and their goals, and the clinician's statement of therapeutic intent. Pretherapeutic assessment should be done at this time to avoid the clinician influence which develops in later interviews.

The data to be collected from each individual should include demographic information. This can be utilized to determine presence or absence of relationships existing between various population groups as characterized by age, race, ethnocultural differences, socioeconomic class, marital status, and family variables. In addition, a sexual profile should be administered that includes items on sexual information, sexual attitudes, sexual behaviors, and sexual anxiety. Such major traits as anxiety, depression, personality disorders, or pathology could be assessed by the use of a well-standardized instrument such as the MMPI. Self-esteem and self-assertion could be measured by behavior scales.

Finally, each individual would be asked to assess his or her unit. This would include the partner's expectation of the individual, the individual's expectation of the partner, and identification of conflict areas. General relationship factors can be assessed using a questionnaire to identify conflicts, unit integrity, and unit strengths. The actual strengths and weaknesses of the unit can be examined in light of the individual's idealized expectations. If a wait period occurs before initiation of treatment, preselected portions of baseline testing should be readministered at the start of therapy, to assess behaviors and anxieties at that time. Evaluations during therapy should be keyed to clinical impressions. The clinical impressions checklist should ideally include the performance of assigned tasks, the reaction to assigned tasks, communication abilities, new behavior assumption, recognition of anxieties and conflicts, and destructive vs. constructive activities. When persistent therapy resistance or previously unknown variables are introduced, objective reassessment of those variables should be done.

THERAPY EVALUATIONS

On termination of therapy, evaluations should be made for both the individuals and the unit to determine the existence of symptoms, whether the stated goal was achieved, the patient's evaluation of sexual function, actual sexual satisfaction, and overall unit relationships. These can be assessed by the patient's self-evaluation and the clinician's evaluation. Follow-up visits can include the previously mentioned clinical checklist and evaluations plus assessments by the units. These impressions should be compared with objective data to evaluate the effectiveness of the treatment and the growth, regression, or maintenance of a unit's success. The use of symptom absence or continued presence, particularly that of presenting complaint, is an important criterion of outcome, but does not completely define therapeutic goals. A major purpose in the follow-up of data collection would be to identify dynamic problems dealing with sexual dysfunction, to develop specific training and therapy modalities along common patterns of the various dysfunctions, to allow greater flexibility in the use of therapeutic personnel, and to establish prophylactic, educative programs.

Group Discussion

Dr. Gebhard questioned the Masters and Johnson definition of success in treating premature ejaculation — namely, the ability of the male to continue coitus until the female achieves orgasm in 50% of encounters. This definition seems to make male premature ejaculation a function of the female partner's orgasmic competence. Another definition was offered by Dr. Green — the suggestion of Dr. Donald Hastings of the University of Minnesota Medical School that a premature ejaculator is a man who cannot continue coitus as long as he wants to. This is a male-based definition; a man who has orgasm when he doesn't want to lacks ejaculatory control. Dr. Rose doubted that a man who could continue coitus for an hour or two should be classified as a premature ejaculator merely because he wanted to continue even longer. He suggested that time to male orgasm must be distributed along a bell-shaped curve; thus males at the extreme short end of the curve might properly be labeled premature ejaculators. Dr. Forney-Settlage replied that in fact the patient defines the problem; if troubled, he or she comes for help. Sometimes the problem is not really a sexual problem; for example, young college women often come in distressed that they have one orgasm instead of a series. This can be diagnosed as a problem of unrealistic expectations and treated as such. Merely talking about this often relieves the distress.

Dr. Rose again asked for specifics: after how many minutes of coitus does a male's problem cease to be premature ejaculation and become a problem of unrealistic expectations? It was suggested that increasing duration beyond a reasonable limit does not in fact increase a women's likelihood of achieving orgasm and that this limit may be surprisingly short. Thus, as a rough rule of thumb, if a man can engage in 10 min of unrestrained foreplay plus 4-6 min of vigorous coitus, increasing these durations is unlikely to improve his partner's orgasmic potential.

Dr. Fordney-Settlage agreed that in particular cases it is necessary to distinguish between premature ejaculation in the male and hypoorgasmia in the female. A case was cited in which the wife complained of her husband's premature ejaculation despite an average of 40 min spent in coitus. Dr. James reported that among medical school students couples were found in which the male was taking on full responsibility for bringing on female orgasm while the female was failing to meet him halfway by shortening her own time needs — an insidious situation.

Cases were also cited in which the female reaches orgasm very quickly and finds further coitus painful; the male in such cases fails to reach orgasm and is left in the frustrated role traditionally assigned to the woman. Perhaps, it was suggested, a treatment comparable to the squeeze technique for the treatment of premature ejaculation should be developed for premature *female* orgasm.

Dr. Fordney-Settlage noted that in her sample there were many cases which did not fit the classic Masters and Johnson categories. Thus 18 women in a particular ethnic minority presented with complaints of anal dyspareunia. They didn't care whether they had orgasm or not, they just didn't want anal intercourse to hurt so much. This she considered a sexual dysfunction which can be treated.

The high incidence of ejaculatory incompetence in one ethnic group was commented on. Dr. Fordney-Settlage replied that this might be a happenstance. All such cases in her sample were referred by an infertility clinic; thus a treatment center without such referrals would see fewer or no cases.

She was also asked whether physical function in her patients was normal. She replied that one group of males was referred from cardiology following myocardial infarcts; physical function was fairly normal but fear levels were high. Reassurance proved helpful. About one-third of the women with dyspareunia had physical conditions which appeared to be contributory — for example, post-infection scarring; some cases required surgical repair followed by the usual sexual training program. Three patients were diabetic; all were treatment successes. One unusual patient alleged that she was allergic to her husband's semen and broke out with hives when exposed to it. Her diagnosis proved correct; no hives followed when a condom was worn.

Dr. Schmidt asked whether follow-up studies could not describe outcomes in more significant detail than merely the success/failure dichotomy. Dr. Fordney-Settlage replied that she hoped to begin much more meaningful post-treatment evaluations when funds became available and has prepared a detailed evaluation outline. (It was circulated to conference participants.)

Dr. Schmidt reported that at his center in Hamburg evaluation methods were being developed in an effort to compare various modes of therapy. His impression was, for example, that in the Hamburg setting therapy with one therapist was about as effective as with two, and much less costly. (Dr. Fordney-Settlage agreed.) A schedule of five daily sessions per week appeared indicated for some kinds of patients; others seemed to do better with two sessions a week for a longer period. Comparisons were also planned between pure Masters and Johnson type therapy and therapy which combined that with a form of communications therapy. Within a year or so he hoped to be able to report comparative results based on full evaluations — including a structured diary-type recording of sexual activity 3 months before therapy, immediately before therapy, immediately after therapy, and 3 months and 1 year after therapy.

Dr. Fordney-Settlage reported that one advantage of the setting in which she worked was that it drew patients from a population receiving ongoing general health care at her center. Thus her population can be used to generate hypotheses which can later be tested; in the absence of an ongoing population, hypotheses become merely guesses.

She also mentioned the use of relaxing drugs in the treatment of a few cases of ejaculatory incompetence. Mr. Brecher called attention to the rarity of such mentions in the literature of sex therapy. She replied that there had been other mentions and isolated case reports, but agreed that there was a dearth of well-controlled studies of the effectiveness of drug therapy for sexual dysfunction.

Dr. Rose mentioned the widespread belief among psychiatrists that patients coming for sex therapy with sexual dysfunction as the presenting symptom are in fact seeking broader help with respect to themselves and their interpersonal relationships. Dr. LoPiccolo replied that while there certainly are such cases there are also many cases where the sexual dysfunction is itself primary. Among patients presenting with premature ejaculation, for example, perhaps 90% are not the result of marital dysfunction but are in fact primarily a sexual problem. There may be a coexisting marital dysfunction, but it is not causing the premature ejaculation. Much the same is true of primary orgasmic dysfunction. Secondary orgasmic dysfunction — that is, dysfunction arising after a period of healthy functioning — and secondary erectile dysfunction are more likely to have roots in marital dysfunction. Dr. Rose called attention to the need for valid data on this issue when gauging the relative merits for psychotherapy and sex therapy. Dr. LoPiccolo noted that where a man ejaculates prematurely with his wife but not with another woman primary marital dysfunction is quite probable.

Dr. Green asked whether sex therapy was casting fresh light on the question of the etiology of sexual dysfunction. Dr. Fordney-Settlage replied that the thorough sexual histories her center was collecting should, when analyzed, indicate what types of prior experiences are associated with a high risk of sexual dysfunction.

Dr. LoPiccolo stressed the need for a control group in assaying sex histories. Thus most anorgasmic women report that they were inadequately prepared for their first menstruation and were in general warned against sex. Much of the same is true, however, of orgasmic women and even of multiorgasmic women; unpreparedness for first menstruation and warnings against sex are modal for our culture.

Dr. Rose suggested that, as a result of the sex therapy movement, people were identifying themselves as having problems which were not in fact problems — as in the case of the woman who complains she is not multiorgasmic. Dr. Forney-Settlage replied that there are also, in contrast, many people in our culture with true sexual dysfunction who fail to recognize it as a problem. Thus out of every 1000 women coming to her center for general medical care and answering a survey questionnaire some 350 do not know what an orgasm is; and on follow-up many fail to recognize that their inability to have orgasm is a problem.

Dr. James called attention to allegations that, as a result of the women's movement, more and more men are feeling inferior and finding themselves impotent. She asked whether Dr. LoPiccolo had observed such an increase in impotence. He answered no; but he had some relevant data. Of women referred to his sex therapy clinic, perhaps 80% phone for an appointment. Among men referred, only 10% phone — indicating a great resistance among men to identifying their sexual problems as psychological. Further, women commonly discuss their problem only with the referring physician. Among men, in contrast, many have long doctor-shopped in search of someone who could give them a pill, a shot, a tonic, or some other somatic cure. Thus the apparent increase in men presenting for impotence may in fact be simply a decline in resistance to sex therapy. Dr. Green hypothesized that as the knowledge that sexual dysfunction is reversible spreads more and more impotent males come for help, thus creating the impression of a rising incidence. Also, men are increasingly coming in with the complaint that they cannot produce orgasm in their partners; they see this as their own failure, and need continuous external validation of their sexual competence.

The discussion concluded with a remark that the site of sex therapy is the therapist's office; thus responsibility for translating office talk into bedroom behavior lies solely with the patients. At least two paramedical services, those supplied by the visiting nurse and the social worker, are sometimes offered in the home. Could these home visits be of use in the delivery of sex therapy services?

Sexuality and Physical Disabilities[1]

Theodore M. Cole, M.D.[2]

INTRODUCTION

Physical disabilities which produce physical handicaps are increasing in frequency. As many as 10% of the general adult population have a physical handicap (Finley, 1972). Types of handicaps include disabilities which have their onset at birth or in early life, those which have an abrupt onset after puberty, disabilities which have progressive courses and begin before puberty, and progressive disabilities which begin after puberty. Examples include arthritis, amputations, deformities, cerebral vascular disease, coronary heart disease, kidney failure and transplant, blindness, deafness, developmental disabilities, paralysis, and disfigurements. In spite of common belief to the contrary, many disabled people report that their disabilities do not alter their sexuality or libido (Richardson, 1972).

The sexuality of physically disabled people can no longer be ignored by the medical community. Weiss and Diamond (1966) reported on a group of adults with paraplegia and found that those who avoided realistic acceptance of their disability were also avoiding a realistic and conscious consideration of their sexuality. The juxtaposition of these two characteristics is important to the rehabilitation effort since rehabilitation depends on the individual's ability to recognize and accept life as it is. Indeed, most rehabilitation professionals have experienced the frustration of working with a patient who is neither able nor willing to reconstruct his life based on reality but rather insists on unrealistic hopes. It is my belief that clinicians who train themselves to be comfortable and capable of working in the area of sexuality with physically handicapped adults may discover that in so doing they may have uncovered a totally new ability to facilitate other areas of reality acceptance.

[1] This paper was presented at the conference, "Sex Research: Future Directions," held at the State University of New York at Stony Brook, Stony Brook, New York, June 5-9, 1974.
[2] Program in Human Sexuality, Physical Disabilities Unit and Department of Physical Medicine and Rehabilitation, University of Minnesota Medical School, Minneapolis, Minnesota 55455.

My colleagues and I have reported on a series of group discussions about sexuality for spinal cord injured adult men and women (Cole *et al.*, 1973). We conducted discussion seminars on human sexuality with medical students and spinal cord injured adults, both single and married. To the surprise of some, we found that the physically disabled adults were less inhibited in their discussions of sexuality than were the able-bodied people and medical students. The seminar format used explicit movies to evoke feelings which served as a basis of discussion. Among the disabled people there was less wasting of time in social chit-chat and there were fewer expressions of boredom, criticism, and annoyance with the seminar content or format. Our experience with this group and others like it has shown that adults with this physical disability are concerned about reproduction and self-image. Fantasy is also important. In spite of the physical limitations of the disability, many paraplegic and quadriplegic people state that sexual activity is personally satisfying and much of the satisfaction is derived from satisfying their partners. The partner's ability to participate in the sexual experience stimulates the disabled person's own pleasure. We also learned that we must give up the belief that physically disabled people cannot or do not wish to frankly discuss sexuality, either their own or that of others.

The situation for the developmentally disabled person is somewhat different. It is axiomatic that the able-bodied child has an assigned sex role accompanied by appropriate urges and that he or she will become socialized through contact with schools, peers, and sharing in sexual experiences. In this manner, the normal child will develop a healthy image as a male or a female which will lead to realistic world views. When the maturation process is complete, the child will take his place as a responsible adult in society. The developmentally handicapped child, on the other hand, leads a sheltered life. For his entire life he may depend on others for self-care. Many of his contacts will be through institutions which will foist the institutional ethic on the child. In many cases, the parents, too, will be unable to accept the child's sexuality as an active and growing component of his maturation. Although the child will receive the same sexual stimuli as normal children, e.g., movies and TV, he is not socialized in a comparable way. He is supervised much or all of the time, with subsequent inhibition of his ability to proceed through the socialization process expected for the normal child. It follows, therefore, that the congenitally handicapped child may become unrealistic in defining his role as a sexual person in a society where role definition depends in large measure on the sexual content and imagery of ordinary daily communication.

Much depends on whether the disability is acquired at birth, prepubertally, or after puberty. Much also depends on whether the disability has its onset suddenly or gradually, and whether it is stable or progressive.

I shall single out one form of handicap — spinal cord injury — for further discussion, since it is one type of injury very likely to give rise to serious sexual dysfunction at the physiological level, yet is consonant with a high degree of psychosexual fulfillment at the psychological and interpersonal levels. It also happens to be the form of physical handicap which has been most studied from the point of view of sexual function and dysfunction.

SPINAL CORD INJURY AND SEXUALITY

It is often difficult for others, as it was for me, to appreciate the obvious and subtle aspects of the physical disability of spinal cord injury. However, I offer a method of facilitating an understanding of the disability for people who have had little or no contact with it. It would be helpful if the reader would imagine that he has become suddenly paraplegic. He is totally paralyzed and has lost all ability to feel bodily sensations from the mid chest down. He is a head and shoulders floating in space. He can still feel normally above the mid-chest area. Automatic bodily functions have been altered and bladder and bowel control has been lost. The bladder may empty itself spontaneously into diapers or clothing, or it may be continually drained off by a tube in the bladder which passes out of the urethra and travels down to a plastic bag strapped to the leg and filled full of urine. Because bowel control has also been lost, fecal incontinence can occur at any time. There is no superficial or deep sensation in the genitals, and one would have to be watching to know that the genitals were being touched or manipulated. The physical experience of orgasm no longer can occur. In the case of the male, erotic stimulation which previously produced erections no longer does. The body has become abruptly altered and its normal contours have been changed. Areas of looseness and sagging may appear together with other areas of atrophy and loss of muscle mass. The individual can no longer stand or walk. He moves about seated in a wheelchair, causing him to converse with erect people by looking upward. Everyone is taller than he. It is not difficult to imagine that this altered body condition could lead to self-consciousness and feelings of inadequacy which may cause the individual to actively avoid sexual encounter. Yet through all this it must be remembered that the capacity for psychosexual enjoyment remains in most people thus disabled.

Male Sexuality and Spinal Cord Injury

Most of the psychological literature is concerned with the premorbid and postmorbid psyche of the individual with spinal cord injury. Sexuality is acknowledged as important, but very few guidelines are provided to the clinician to assist in working with the patient.

Most of the medical literature on spinal cord injury tends to be mechanical and emphasizes penile-vaginal intercourse or mechanical genital stimulation.

Zeitlin found that 54% of 100 spinal cord injured men were capable of achieving penile erections and 26% could engage in penile-vaginal intercourse. However, only 5% reported the sensation of orgasm and only 3% reported external ejaculations (Zeitlin et al., 1957). In Japan a larger group of 655 paraplegics and quadriplegics were studied by Tsuji et al. (1961). Slightly over half of the male patients were able to have penile erections at the time of discharge from the hospital and over three-fourths were reporting penile erections 1 year after injury. In the 150 patients he studied, Comarr (1970) reported that psychogenically stimulated erections occurred in about 25%. Three-fourths reported erections occurring spontaneously, due to either external or internal stimulation. Mechanical stimulation could achieve erections in approximately 70%. However, only 38% reported that they had attempted to engage in sexual intercourse and only 24% of those stated that they were successful at it. Poignantly, 38% reported that they had not even attempted coitus.

In the able-bodied male, erection of the penis occurs as a result of psychogenic stimulation while awake or in association with periods of rapid eye movement during sleep. In the adult with spinal cord injury, erections more commonly occur as a result of reflex activation of the blood supply to the penis. Reflex erection may be brought about by external stimulation applied to the body somewhere below the level of the spinal cord injury. Primary sites for effective stimulation are found in pelvic organs. Depending on the individual and the situation, the effective stimulus may be touch or noxious stimuli such as anal manipulation, squeezing of the glans penis, or pulling of pubic hair. Spontaneous penile erections may result from visceral stimulation such as bladder distension or fecal compaction in the rectum.

As noted earlier, fewer than 5% of adult males with complete spinal cord transection are capable of external ejaculation. However, some of the men who do not report it do report a perception of heightened spasticity before and/or at the point where ejaculation would have occurred. Many individuals with somatic spasticity, from either brain or spinal cord injury, report that generalized reduction in spasticity may follow the completion of intense sexual activity and that it may last for several hours. If the spinal cord has been injured in the lower sacral segments, a flaccid paraplegia may ensue due to injury to the final common pathway. However, 20% of such patients report external ejaculation even though they lack sensation in the genitals (Bors and Comarr, 1960).

Fantasied orgasms or orgasms "in the head" are reported by a sufficiently large number of spinal injured adults, both male and female, to make it an important inclusion in this report. Furthermore, it is difficult for many able-bodied people to understand a fantasied orgasm or to believe that it could be sufficiently satisfying to the paralyzed person. However difficult it may be to understand, many spinal injured men and women report orgasms in spite of complete denervation of all pelvic structures. The orgasm so reported often leads to a com-

Table I. Comparison of Sexual Response Cycles in Able-Bodied and Spinal Cord Injured Males: Male Sexual Response Cycle[a]

	Able-bodied male	Spinal cord injured male
Penis	Erects	Erects
Skin of scrotum	Tenses	Tenses
Testes	Elevate in scrotum	Elevate in scrotum
Emission	Yes	No
Ejaculation	Yes	No
Nipples	Erect	Erect
Muscles	Tense, spasms	Tense, spasms
Breathing rate	Increases	Increases
Pulse	Increases	Increases
Blood pressure	Increases	Increases
Skin of trunk, neck, face	Sex flush	Sex flush

[a]Reprinted, with permission, from *Human Sexuality: A Health Practitioner's Text,*
copyright 1975, The Williams and Wilkins Company, Baltimore.

fortable resolution which seems similar to that experienced by neurologically intact individuals. Some paralyzed people report they have the ability to focus on
sensations being received from portions of their body still innervated, and, by
concentration, to enhance that stimulation and transpose it to a part of the body
that is anesthetic, such as the genitals. Using this technique, some spinal injured
men and women report orgasms not only once but many times during a single
sexual encounter.

The sexual responses of which the completely spinal cord transected male
is capable are, surprisingly, not very different from those of neurologically intact
males (Table I). This is not to minimize or deny the dramatic impact of paraplegia on sexuality but rather to put the alterations into perspective. Penile erection can still occur, although not as predictably as before, nor is it still produced
by erotic stimulation. Emission and external ejaculation usually do not occur in
spinal injured men and therefore the external evidences of sexual "success" are
not seen. I have found it to be of some clinical comfort to paraplegic men to
point out that their ability to respond sexually is not greatly different from their
physiological capacities prior to injury. Providing this information is in keeping
with the notion that patients do better when they are given as much information
as possible about their health and altered body state.

Female Sexuality and Spinal Cord Injury

Four out of five spinal injuries sustained by adults in the United States
today are suffered by men. Well over 80% of the literature on spinal cord injury
deals with male sexuality. However, lately some information about female sexuality and spinal cord injury has become available and much of it emphasizes

that the female's sexuality may be less threatened by her spinal cord injury. Reasons for this lie in the differences in physiology, anatomy, and cultural expectations for the two sexes.

Paralysis and the accompanying loss of muscle strength may more easily conform to society's view of the female's sexuality, which in our culture is still thought of as more passive than the male's. Women might vigorously argue that physical limitations are as restricting on their sexuality as on a male's. However, the medical literature does stress this difference, also emphasizing that women will generally find their sexual options less limited by paralysis than will men. A cardinal example of this is reproductive fertility following spinal cord injury. The female's reproductive capacity is virtually unimpaired, whereas the male's is greatly reduced or eradicated.

Women may experience other changes following spinal cord injury. Loss of menstrual periods will be experienced for up to 6 months after the injury by approximately half of spinal cord injured women. However, normal menstrual cycles will reoccur in almost all women within 1 year. Painful menstruation or pain on intercourse, if present prior to injury, will be absent because of loss of prior ability to sense pain. Just as for men, the ability to sense physical orgasm may be lost, but the woman, too, is capable of experiencing sexual arousal, plateau, orgasm, and resolution. She may also experience fantasied orgasms just as do men. As in males, those areas of the body which retain neurological innervation may become more highly eroticized than they were before the neurological injury.

With the return of regular menstrual periods, fertility is restored and pregnancy becomes as possible as it was prior to injury. The spinal cord injured woman is capable of carrying a fetus to term without serious medical problems. Most deliveries are from below and caesarian section is no more common among spinal cord injured women than among their able-bodied counterparts.

Since fertility is unaffected, contraception is important for the clinician to consider and provide. The increased incidence of venous thrombosis associated with oral contraceptives is enhanced by the propensity for venous stasis and thrombosis in the paralyzed limbs. The physician should also be aware of potential problems associated with intrauterine contraceptive devices in a woman whose ability to sense pain is impaired. A woman whose hands or upper extremities are paralyzed by spinal cord injury may have difficulty inserting a contraceptive vaginal diaphragm without the aid of her partner.

As a result of ignorance and ill-founded attitudes, many physicians recommend that spinal injured women be permanently sterilized and not raise a family. However, our experience has been contrary to that advice. We have worked with many women who manage small children, household duties, and careers from their wheelchairs when they are supplemented with the supportive assistance they need to deal with their disabilities. Physicians would be advised to

Table II. Comparison of Sexual Response Cycles in Able-Bodied and Spinal Cord Injured Females: Female Sexual Response Cycle[a]

	Able-bodied female	Spinal cord injured female
Wall of vagina	Moistens	−
Clitoris	Swells	Swells
Labia	Swell and open	Swell
Uterus	Contracts	−
Inner two-thirds of vagina	Expands	−
Outer one-third of vagina	Contracts	−
Nipples	Erect	Erect
Muscles	Tense, spasms	Tense, spasms
Breasts	Swell	Swell
Breathing	Increases	Increases
Pulse	Increases	Increases
Blood pressure	Increases	Increases
Skin of trunk, neck, face	Sex flush	Sex flush

[a]Reprinted, with permission, from *Human Sexuality: A Health Practitioner's Text,* copyright 1975, The Williams and Wilkins Company, Baltimore.

individualize their advice to spinal injured women and base it on the couple's wishes and their ability to adjust in an overall fashion to the handicaps associated with spinal cord injury.

Although less is known about the sexual responses of paraplegic and quadriplegic females, some information is available in the literature and we have collected more anecdotally and clinically (Table II). As for the male, the woman's sexual responses after spinal cord injury are more noted for their similarities to than their differences from those of neurologically intact women. She too is benefited from instruction in the nature of her disability which may give her coping skills.

SOME KEY POINTS

As one might expect, many of the problems faced by spinal injured adults have direct counterparts in the able-bodied population: comfort, confidence, competence, and relationships. Their ability to communicate wants and feelings with their partners, their mutual willingness to experiment with sexual activities which are pleasing and not exploitive, emphasis on fantasy, a reasonable program of physical hygiene, and the knowledge that more sexuality lies within the head than between the thighs all help to set the stage for restoration of an active and satisfactory sex life.

Sexual options for spinal injured adults should be as available and free from limitations of ignorance, fear, and guilt as for able-bodied people. This advice applies equally to the adult who elects sexual indulgence and to the one who elects celibacy. It should be clear to the patient, the family, and the professional that a satisfactory sex life is possible. The rehabilitation hospital should be expected to teach patients about sexual behaviors in the same way they teach other activities in daily living, such as self-care, mobility, and communication.

How should the physician approach the spinal injured male with an injury in the lower sacral segments, producing permanent and irrevocable flaccidity of the penis? This situation poses a special problem for the professional since we know how devastating impotence can be to an able-bodied man. We have been teaching couples the technique of "stuffing," a technique which is included in the sexual therapy sometimes utilized in the treatment of premature ejaculation. For those couples who wish to experience intravaginal containment of the penis, the stuffing technique provides them with an option which might otherwise seem unattainable. In some cases where religious beliefs dictate the necessity of penile-vaginal intercourse, stuffing can mean the difference between endorsement and nonendorsement of the marriage by the church. With cooperation and experimentation, the paralyzed male can take a position on top or underneath the partner and the flaccid penis can be "stuffed" into her vagina. By voluntarily contracting her pubococcygeus muscles around the penis, she can hold it within the vagina and even create a "tourniquet" effect, causing the penis to become semierect.

I should like to be able to present a similar review of sexual function and dysfunction in other physical disabilities. Even if time were available, however, I could not — for many of the data are not available. Learning at least as much about sexual function in the other forms of disability as is already known about spinal cord injury — and ideally learning much more — should have a high research priority.

COUNSELING

The physician should carefully consider the wide assortment of methods by which patients and hospital staff deal with sexual concerns. Avoidance with or without embarrassment is the most common. Where avoidance is not practiced, the staff may feel comfortable discussing limited aspects of sexuality such as fertility and pregnancy. Although sexual motivations are myriad (Neubeck, 1972), it is less common for the hospital staff to discuss sexuality in terms of pleasure, recreation, or communication.

In our experience, it is rare for the physician and the patient to discuss sexual activities other than penile-vaginal intercourse, such as pleasuring techniques, oral-genital stimulation, and the use of devices such as vibrators. It is extremely rare for the physician to instruct the patient in the use of fantasy. When the patient's sexual activity is homosexual or otherwise variant, physician-patient communication is ordinarily further restricted.

If the hospital staff has carefully laid the foundation for sexuality to be included with other health issues, the paraplegic may present a host of questions which challenge the staff's clinical acumen as well as their personal attitudes. The patient may ask if intercourse is possible while a urethral catheter is still in the bladder, or how can sexual activity be carried out when there is a risk of bladder or howel incontinence? Does the presence of an ileal conduit prohibit sexual activity (an operation which diverts urine flow away from the bladder to a loop of intestine which exits it at the abdominal wall into a plastic bag)? The patient may experience and ask about medical problems, such as severe headache, caused by stimulation of the genitals. Some may be concerned about the partner's developing a secondary infection by having sexual intercourse while the urine is infected, or the couple may ask if they can remove the enlying urethral catheter for purposes of sexual activity. Other questions having more to do with sexual technique may include queries about physical positioning for purposes of sexual intercourse and methods for having "quickies" or intercourse with only a brief amount of time available, precluding transferring to a bed and undressing. People may ask how to have sex without getting out of the wheelchair.

We have found that many of these questions can be answered with little more than open communication on the part of the patient and the partner and a willingness to experiment. For example, a woman can position her catheter along the thigh and tape it there. A man could fold the catheter over the glans penis and direct it along the shaft in order to accomplish penile insertion. The male also may choose to pull a condom over the penis and the catheter together and lubricate it, knowing that the vagina is quite capable of accommodating the added bulk of the catheter. Those who experience sphincter incontinence during sexual activity find that the best way to deal with this is adequate communication with the partner prior to beginning sexual activity and then avoidance of those physical activities which their experience tells them precipitate emptying of the bladder or bowel. The plastic bag which collects urine from the ileal conduit can simply be positioned out of the way and the couple can avoid chafing of the bag, which may cause leakage around the seal. Many people keep a towel or two handy in case of accidents. A quadriplegic's headache caused by genital stimulation is the result of autonomic hyperreflexia, with blood vessel spasm and secondary elevation of blood pressure. This well-known medical phenomenon is evoked reflexly, most often by stimulation of pelvic organs. Headache can be quickly relieved by stopping the stimulation that evoked it, resuming activity when the discomfort has passed, and taking care not to duplicate the stimulation. Brief sexual encounters can be experienced while sitting in a wheelchair and often utilize oral-genital activity. For intercourse, the disabled male may position his female partner on his lap with her back toward him or he may remove the arms of the wheelchair and she can sit astride his lap facing him. Urinary infections will almost never cause a significant infection of the partner when sensible hygiene is observed. The urethral catheter can be removed prior to sexual activity and a sterile one reinserted when sexual activity is concluded, provided that the couple does not unduly prolong the period without the catheter and thereby cause overdistension of the bladder.

The rehabilitation center at the University of Minnesota has been providing sexual counseling on a regular basis to all patients who wish it. However, preparation of the hospital staff was necessary before this could take place. The Sexual Attitude Reassessment (SAR) and Counseling Seminar originally developed by the National Sex Forum, San Francisco, California, and modified by the University of Minnesota's Program in Human Sexuality was offered to all staff. Further modifications in the model were made to meet specific needs of the rehabilitation center. The result of this deliberate approach was that sexuality took on a respectable aspect for both patients and staff. It soon became apparent that sexual function would be included among other health issues and be dealt with openly in a matter-of-fact manner. Patients and their families came to realize that treatment of sexual problems associated with physical disability could be expected and questions could be asked of anyone on the rehabilitation team.

We involve the patient's partner in the counseling process whenever possible. However, some physically disabling conditions occur in people who have no sexual partner. To require these counselees to participate only in couple counseling would be inappropriate. Counseling techniques are designed to fit the needs of the single individual or the couple as needed. However, our experience continues to reinforce earlier work of others that sexual counseling is greatly helped if the counselee has a sexual partner with whom to work on the problems and solutions developed in therapy.

We have found it unnecessary and unwise to defer raising the subject of sexuality or to wait for the patient to ask questions. The dramatic disability and its correlates may be so overwhelming that the patient and the family may not think of sex or feel that it is permissible to ask questions. The professional is best advised to observe the medical model in this situation and to anticipate the need and provide information as appropriate.

Here are some general guidelines for talking with the recently physically disabled patient in the hospital: Don't be frightened or put off by an initial negative response from a patient. Disabled people are not fragile and can process sexual information at least as well as able-bodied people. Conduct the discussion in an understanding and sincere manner using eye and hand contact in order to extend comfort. Above all, make it clear that sexual feelings are natural and expected. Discussion of sexuality should not be segregated from other aspects of medical interviewing, and is best placed in the context of the variety of problems which the patient faces: medical, social, psychological, and vocational. We have found that most patients are interested and able to talk about the impact of their disability on their sexuality as soon as the life-threatening aspects of the disability have passed and there has been enough time to contemplate life after discharge from the hospital. A complete discussion should include physical, occupational, personal, and recreational aspects of the disability. Through counseling, past living patterns should be discussed and the patient and partner should begin to anticipate changes that the future may hold. These changes in performance capabilities lead easily into discussions of sexuality. "Can I masturbate?" "What are my concerns about fertility?" "Can I consider myself an acceptable marriage partner?" "Can anyone like my body?"

In the initial interview it is fitting to ask questions about the patient's partner. Not only may the partner have questions, but also discussion of the partner's concerns may help the patient to focus on important aspects of rehabilitation. For some time after the onset of the disability, the patient may not fully appreciate the importance of communication for physically immobilized people. However, it will eventually become apparent that communication skills can manipulate and control the environment or attract attention, thus compensating for lost mobility.

FUTURE DIRECTIONS

One day it will become necessary to provide evidence that sexual counseling for physically disabled people is worth the effort. In most cases, the physical disability is so severe that insurance or public funds will be required to pay medical and rehabilitation expenses. If additional expenses for sexual counseling are to be underwritten, those who set policy for insurance or public funds will want to know if a benefit can be anticipated.

It will also become necessary to better understand the relationship among self-esteem, sexuality, and disability. A positive self-esteem may correlate with the patient's ability to develop compensatory mechanisms with which to reengage the world. Figure 1 illustrates how self-esteem relates to work on the one hand and sex on the other. If one assumes that successful work leads to increased self-esteem, then it is reasonable to assume that one will continue in a work activity which has produced increased self-esteem. Those activities will benefit self and probably will lead to a decreased willingness to accept the dependent state. In the case of the physically disabled person, this will lead to a decrease in his complaints and need for outside medical support.

In a parallel sexual model, increased sexual success leading to increased self-esteem will encourage the individual to continue to engage in those activities which have increased the self-esteem. These activities will not only satisfy self but will probably also decrease feelings of castration which may otherwise result from the disability. With reduced feelings of castration, there will be less need for social withdrawal and therefore less need for outside social support. The result will be a reduced social and dollar cost to society.

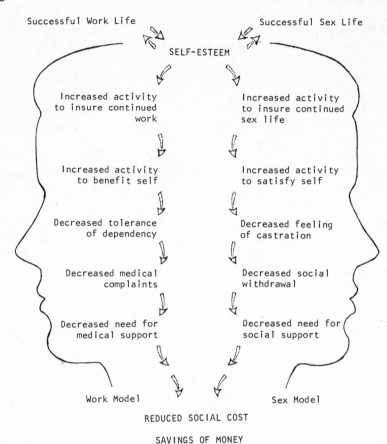

Fig. 1. Relationship of work, sexuality, and self-esteem to cost in the physically disabled adult.

Both of these models suggest that activities deliberately designed to increase the patient's self-esteem may produce decreased costs for the health care system and for society in general. Research needs to be done to explore this thesis so that society will be willing to underwrite activities to produce a greater likelihood for a more satisfactory sex life.

The physiological alterations produced by physical disability are still poorly understood with respect to sexual function. Issues such as fertility and alterations in the normal sexual response cycle need to be studied for all major disabilities.

Ignorance, fear, and guilt continue to adversely affect sexuality and physical disability. Research should be done to explore how one influences the other. Society thinks of the disabled person as infantile and desexualizes him. Research should be done to explore the thesis that bestowing sexual competence will remove feelings of inadequacy for disabled people, thereby restoring self-confidence, self-determination, and self-responsibility. The same reasoning suggests that family disruptions, which often follow unexpected physical disability, may be gainfully influenced by patient and family education in sexuality. At its foundations, rehabilitation is helping the individual regain control over his life. Any modality which helps the patient become self-actualizing and adultlike rather than dependent and childlike may significantly benefit the rehabilitation outcome.

Developmental disabilities are often attended by maturation arrest, produced in part by society's segregation of conspicuously disabled people. Rehabilitation in this group of patients hinges on the individual's ability to develop the social skills to cope in a society which discriminates against people who have grown up obviously different. Now that sexuality and physical disability are becoming more acceptable, research should be done to clarify the role of sexual therapy in training for living. Is it possible that regression, dependency, and lack of self-responsibility would be benefited by training in sexual comfort, confidence, and competence?

Group Discussion

Dr. Lipman-Blumen called attention to certain parallels between the status of the physically handicapped and the status of women in our society. Thus women are deemed weaker, incompetent, and lacking in various valued male characteristics (including lack of a penis). Like the disabled, they are required to exhibit adaptability and plasticity. They are kept dependent on others. Their self-esteem is damaged, as in the case of physical handicap. Like the physically handicapped, women are unnecessarily barred from many jobs and roles. Thus femininity can be viewed as a kind of physical handicap. More important, the methods women are currently using to improve this traditional status might also be used in the liberation of the physically handicapped. Dr. Cole agreed.

Dr. Rose asked whether biofeedback techniques might be useful in attempting to bridge the gap between central and peripheral nervous systems following spinal transection. Dr. Cole replied that this is a possible area for research. He regretted, however, a tendency to stake hope on major technological breakthroughs which might improve physiological function at the expense of efforts to help the physically handicapped make the most of their remaining capacities.

Dr. Lewis questioned whether our concepts of health may not be too culturally limited. Some physically disabled persons, for example, simply withdraw from life. We see this as unhealthy — but might it not be the healthiest possible response? Might we not even encourage withdrawal in some circumstances? Dr. Geiger said he viewed health as the ability to exercise options. If a parent chooses withdrawal from among the options available to him, fine; that can be a health response, but if he withdraws because he sees no alternatives he is sick and needs help. It was also pointed out that by refusing to discuss sexuality with the handicapped or to help them achieve their sexual potential most medical centers are in fact both counseling and encouraging withdrawal. The centers where sex counseling and therapy are available are a handful of exceptions to the general rule.

Dr. Cole suggested that if a hospitalized patient develops a fever which goes untreated and leads to long-lasting damage, that is culpable negligence. The same view might be taken with respect to a patient brought to a hospital for trauma or illness who loses sexual function and zest for life because sex counseling and sex therapy are withheld.

The importance of providing sex education, sex counseling, and, if necessary, sex therapy for the mentally retarded was also commented on.

Dr. Cole suggested that the issue is one of priorities. If a man or women loses a leg, there are facilities at the hospital to treat the acute effects, facilities to supply and fit an artificial leg, facilities to teach the use of the new leg, and facilities to teach the ways in which the activities of everyday life can be carried out with the new leg — but no facilities to teach the exercise of sexuality with the new leg. Is this merely because there are no facilities left over? On the contrary, it is because we place so low a priority on the maintenance of sexual function. The time for a change in medical and hospital priorities has come — especially since it can be shown that maintenance or restoration of sexual function favorably affects other rehabilitation goals and hence tends to lessen overall costs.

Dr. Lipman-Blumen asked about the maintenance of sexual function in those *temporarily* disabled. Dr. Cole replied that one often sees an older couple, affectionate and sexually active, separated when one member is hospitalized. Sexual communication between them is thereby cut off; both the patient and the relationship deteriorate. All that is needed to prevent this deterioration is privacy, which can and should be supplied. Dr. Gebhard noted that the same cruel lack of privacy characterizes homes for the aged, halfway houses, nursing homes, and other institutions in our culture.

Attention was called to the need for *preventive* sex counseling prior to hysterectomy, prostatectomy, and perhaps other surgery.

A research design was discussed in which some patients in a handicapped group — paraplegics, for example — would receive no sex counseling or therapy, some would receive therapy after rehabilitation is otherwise complete, and some would receive routine preventive counseling as soon as possible after injury, plus later therapy if necessary. The effects would be evaluated not only in terms of sexual function but also in terms of length of hospitalization, ability to resume employment, general level of human functioning, total cost of care, and other parameters. Such a study might be based on existing data — comparing results at centers where sex counseling and therapy are routinely offered with results at centers where sexual matters are still taboo.

Dr. Fordney-Settlage called attention to the need for further research as well as sex counseling and sex therapy for patients with diabetes, heart disease, and other medical conditions. Renal disease was mentioned as a special case, since sexually significant hormones are washed out during dialysis.

Dr. James stressed the need to correct mistaken public images of the handicapped as asexual. Thus women married to spinally injured males are often pitied by their friends: "You poor dear." Some reply, "I never had it so good." As this becomes understood, the sexual opportunities of the handicapped will expand. Dr. James noted also that the ability to give joy is itself important for self-esteem and for enjoyment. The physically handicapped person, even when unable to enjoy orgasm personally, may experience an enormous lift when he or she learns that sexual satisfaction of a partner is still possible.

The sexual sequelae of a physical handicap, she continued, are of two types — one the direct result of the disablement, the other the result of social disenfranchisement. It is important to distinguish these, and to repair both types of sequelae as fully as possible.

Much the same is true, Dr. Cole noted, for the aged who suffer sexual disenfranchisement.

Neuroendocrinology: Animal Models and Problems of Human Sexuality[1]

Robert W. Goy, Ph.D.,[2] and David A. Goldfoot, Ph.D.[3]

ANIMAL MODELS OF BISEXUALITY

The hypothesis has been advanced (Phoenix *et al.*, 1959) that hormones present during early stages of development can determine the pattern of sexual behavior displayed by an individual as an adult. The basic position is that during a very restricted period of development (fetal in some mammals such as the guinea pig and monkey, larval in others such as the mouse and rat born incompletely differentiated) secretions from the XY gonad produce changes not only in the gonaducts and external genitalia, but also in the neural tissues mediating sexual behavior. For sexual behavior, at least two distinct behavioral systems are affected: (1) there is a facilitation or an augmentation of sexual responses normally characteristic of the genetic male, and (2) there is an inhibition or suppression of sexual responses normally characteristic of the genetic female.

Data from several laboratories are difficult to reconcile with this hypothesis. First, many species display bisexuality during mating, in that females mount available partners or males accept the mounts of partners and display lordosis or presenting postures. Second, experiments have been able to augment mounting potentials without suppressing lordosis, and, to a lesser extent, they have suppressed lordosis without augmenting mounting. The hypothesis as originally stated does not cover these conditions of bisexuality which are seen throughout the mammalian species, and in fact could be interpreted as a statement of the ideal case that mounting be expressed only by males, lordosis only by females (Beach, 1968). This paper represents an attempt to reformulate the original

[1] This paper was presented at the conference, "Sex Research: Future Directions," held at the State University of New York at Stony Brook, Stony Brook, New York, June 5-9, 1974.
[2] Director, Wisconsin Regional Primate Research Center, and Professor, Department of Psychology, University of Wisconsin, Madison, Wisconsin.
[3] Assistant Scientist, Wisconsin Regional Primate Research Center, Madison, Wisconsin 53706.

hypothesis of the origin of sexual dimorphism, taking into account the new data collected during the last decade and a half. The discussion will be limited strictly to mounting behavior and the receptive behaviors of lordosis in rodents and presenting in the bitch and the rhesus monkey.

Researchers from several laboratories have now demonstrated that the ideal condition of complete dimorphism does not exist for any species studied. Many years ago, Beach (1942) described normally behaving male rats which mounted at high frequencies and which also displayed lordosis when suitably stimulated. In our own laboratory, lordosis was shown to be a common response of newborn normal male guinea pigs (Goy *et al.*, 1967), although lordosis in adult males is very limited (Phoenix *et al.*, 1959). For at least one strain of rat, Whalen and Edwards (1967) have shown that females mounted as frequently as males when injected with testosterone propionate and also displayed lordosis when suitably treated and tested. A comparable bisexuality has been shown for female mice (Edwards and Burge, 1971b) and genetic female guinea pigs (Young, 1969; Phoenix *et al.*, 1959; Goy *et al.*, 1967). Moreover, the normal female guinea pig displays mounting at the time of both spontaneous and induced estrus (Young and Rundlett, 1939; Goy and Young, 1957), and a similar finding has been reported for the female dog (Beach *et al.*, 1972).

More recently, information has been provided on hamsters from a number of laboratories (Swanson and Crossley, 1971; Tiefer and Johnson, 1971; Eaton, 1970). In this species, a limited expression of lordosis in normal males can be regularly and easily induced. The quantitative studies of lordosis in the male hamster carried out by Noble (1973) show that the expression of the behavior is not as good in males that are allowed to complete sexual differentiation as in males that are not. Nevertheless, the expression of lordosis is more complete in normal male hamsters than in normal male guinea pigs and rats. Thus, for the male hamster, bisexuality exists and sexual dimorphism for lordosis is incomplete.

The behavior of male rhesus monkeys parallels that of male hamsters, with the interesting difference that neither mounting nor presenting (the sex response analogous to lordosis in rodents) depends on hormonal stimulation at the time of its expression. Nevertheless, male rhesus monkeys, both intact and castrated, display both mounting and presenting responses, and a marked degree of bisexuality exists in males of this species (Goy, 1968; Goy and Phoenix, 1971). It could be said, in fact, that bisexuality is more completely expressed in male rhesus monkeys than in males of any other species studied so far in the laboratory and possibly in the natural habitat as well. In his early studies of rhesus social groups, Carpenter (1942) remarked on this subject in discussions of homosexual behavior, and the degree of completeness of the feminine repertoire in adult males has been striking to many observers.

The observation of bisexuality in various mammalian species is not new, but from the information currently available a new relation is discernible. Young (1961), in a review of the information available at that time, stated the general conclusion that for mammals bisexuality was more common among females than among males. We have no evidence indicating that this is not the case when all mammalian species are surveyed. However, from the data which have become available in the last 10 years it is also possible to conclude that bisexuality is not expressed equally by both sexes in a given species. In fact, there appears to be an inverse relation between the sexes with respect to bisexuality; thus, for a given species, the greater the bisexuality of the male, the less the bisexuality of the female, and *vice versa*.

Examples of this are not numerous because the number of species investigated in the manner required to reach such a conclusion is still small. In our own work, however, the relationship is clear. Among rhesus monkeys, males show a conspicuous bisexuality throughout early development (Goy, 1968) and to some extent even into adulthood (Carpenter, 1942). Among female rhesus, on the other hand, male behavior is rare or infrequent during the first years of life (Goy and Phoenix, 1971), and there is no evidence that it can be brought to expression in nonmounting females in adulthood even with large doses of testosterone propionate (Goy and Resko, 1972; Eaton *et al.*, 1973).

Our work with guinea pigs illustrates the opposite relationship between the sexes. Male guinea pigs show little or no bisexuality, and lordosis is difficult to discern in adulthood even after treatment with large amounts of estradiol benzoate and progesterone. Female guinea pigs, in contrast, display lordosis and mounting under a variety of endocrine conditions. The relationship for the guinea pig, in which the female displays more bisexuality and the male less, is duplicated in several other mammalian species: rat (Pfaff, 1970; Pfaff and Zigmond, 1971), dog (Beach *et al.*, 1972; Beach, 1970), mouse (Edwards and Burge, 1971a,b). Isofar as we are aware, however, the relationship between the sexes represented by the monkey has been reported in only one other species. With the exception of work reported by Ciaccio and Lisk (1971), other laboratories have consistently found a high degree of bisexuality among male hamsters and a very low degree among females (Swanson and Crossley, 1971; Paup *et al.*, 1972).

The difference between species with respect to whether the male or female displays the greater bisexuality is found as well in different genetic strains of the same species. Most of our published work on the dimorphic expression of mounting and lordosis in guinea pigs has been with those of the Topeka stock. In this stock, which is not inbred, males display little or no bisexuality and females a moderately high degree, but with considerable individual variation, as would be expected (Phoenix *et al.*, 1959; Goy *et al.*, 1964). We have shown in previous publications that female guinea pigs from strain 2 show little or no bisexuality (Goy and Young, 1957) and display mounting behavior infrequently. In addition, we have recently obtained data which are still in preliminary form but which indicate the presence of a fairly marked bisexuality of behavior in males of this strain. Thus the differences in bisexual expression which exist for guinea pigs are identical to the differences which exist across species; namely, the greater the expression of bisexuality for one sex, the less the expression for the other, but the sex displaying the greater degree of bisexuality may vary with the strain.

ORIGINS OF SPECIES DIFFERENCES IN THE COMPLETENESS OF SEXUAL DIMORPHISMS FOR MOUNTING AND LORDOSIS

Once it is recognized that an inverse relationship for bisexual potential exists between the sexes within a species, and that the ideal case of complete sexual dimorphism is not to be found except perhaps for the mythical ramstergig (Beach, 1971), then the task of reconciling these apparent discrepancies with the original organizational hypothesis can be undertaken. We propose that the inverse relationship for bisexual expression by the two sexes can be understood on the basis of the prenatal or larval hormonal conditions which contribute to bisexuality for each sex. The inverse relationship suggests that the hormonal conditions which contribute to bisexuality in the female act to prevent or limit bisexual expression in the male, and the converse is also true.

This position necessarily deviates from an older view that the female of any species represents the "anhormonal" and undifferentiated state, the condition of sexuality that develops in the complete absence of hormonal stimulation during early periods of differentiation. According to that view, if a female from a given species displays mounting, the assumption could be made that this degree of bisexuality could occur without early hormonal action. Correspondingly, one older view of bisexuality in normal males (Young, 1961) held that lordosis could be expressed, but rarely and only under conditions of unphysiological amounts of hormonal stimulation, unusual forms of environmental stimulation, or both combined. We regard the expression of bisexuality by normal males and females entirely differently and see it as relatable to the specific parameters of endocrine stimulation in early stages of development. Accordingly, the differences between species or between genetic strains in the degree of dimorphism for mounting and lordosis can have the same origins as those for bisexuality in each sex.

ORIGINS OF BISEXUAL BEHAVIOR IN THE FEMALE

Most investigators interested in sexual behavior have been concerned with the problem of bisexuality. Beach (1968), in his chapter on mounting behavior of female mammals, came to the conclusion that the neural tissues essential for the display of this behavior were ubiquitously represented among females of all species, and that the causes for differences among species in degree of expression were to be found in the hormonal and stimulus conditions essential for the evocation of the behavior in adults. His article did not speculate on the biological origins of mounting behavior in females, but one could assume that his position on the matter is not greatly different from that articulated by Whalen (1971). Essentially, the position adopted by Whalen states that mounting behavior is an inherent capability of the female rat (and probably other female mammals), and that no hormonal conditions during early development are essential to its later expression. In Whalen's experiments, normal female rats show high frequencies of mounting under suitable conditions, and no endocrine treatment which he has given to the developing female rat has increased the expression of mounting behavior in adulthood.

An alternative to Whalen's hypothesis regarding the origins of bisexuality in the female rat can be formulated from recent experiments. This hypothesis is that the bisexuality of females is a result of endocrine conditions prevailing during a specific stage of early development. Ward and Renz (1972) and Stewart *et al.* (1971), working on the assumption that androgens were the steroids most relevant to development of the potential for mounting, showed that treatments of female rats perinatally with an antiandrogen (cyproterone) resulted in a significant reduction of bisexuality for the female, although the effect was largely limited to a reduced sensitivity to testosterone propionate administered in adulthood. Females treated perinatally with antiandrogen required more testosterone propionate in adulthood than normal females to display mounting behavior. Ward and Renz appear to have obtained stronger effects than Stewart *et al.*, perhaps because the acetate rather than free cyproterone was used. In addition, the report by Ward and Renz demonstrated that with a specific prenatal treatment lordosis was relatively unaffected but reduction of mounting was marked.

An additional experiment compatible with the hypothesis that early endocrine conditions determine the degree of bisexuality in the female is that of Clemens and Coniglio (1971). They found that the amount of male behavior displayed by adult female rats was directly related to (1) the number of male siblings in the litter and (2) the proximity of the female to a male fetus *in utero*. The data of Clemens and Coniglio are compatible with the view that the origins of mounting in the female rat exist in part in the exposure *in utero* to the relevant steroids which determine the development of mounting in the male siblings. Such an interpretation would have to postulate transplacental transfer of the relevant steroids from the male to the female fetus. If this is in fact the case, then the origins of bisexuality in the female rat might be unique, either to the rat or to polytocous species, and the mechanism could not account for bisexuality in species which ordinarily give birth to only one offspring. The comparative data show, however, that mounting behavior is displayed by females of monotocous species such as the cow and ewe (Young, 1961). In addition, the mechanism cannot adequately account for strain differences of bisexuality of the female which exist in polytocous species such as the guinea pig. Strain 2 female guinea pigs, which show much less mounting than Topeka females, do not differ from the

Topeka stock in terms of litter size or the average number of male siblings per litter. A preliminary evaluation of the effect of male siblings on the mounting of female guinea pigs from a genetically heterogeneous stock (Topeka) indicates that no significant relationship exists (Goy and Bridson, preliminary findings). To be consistent with the new hypothesis, it is proposed that appropriate steroidal exposure must occur *in utero* for these species via ovarian, adrenal, or placental sources. The steroid moiety and/or the temporal parameters of stimulation would necessarily be such as to have little or no effect or lordotic mechanisms or genital structure, while they would serve to augment mounting.

ORIGINS OF BISEXUALITY IN THE GENETIC MALE

Hypotheses regarding the bisexuality of males in various mammalian species have not been as diverse as those formulated regarding bisexuality of females. This may be because, as Young pointed out, bisexuality is more common among females, especially in those species usually studied in the laboratory. There has been a tendency on the part of most investigators to view the bisexuality of males as (1) inherent, (2) limited, and (3) directly controlled by the amount of steroid present in critical stages of early development. It was a surprise to us and probably to many other investigators when the first reports from Swanson's laboratory (Swanson and Crossley, 1971) and from the work of Tiefer and Johnson (1971) showed that normally differentiated male hamsters could easily be induced to display lordosis by suitable treatment with estradiol benzoate and progesterone in adulthood. No investigator has postulated directly that the display of lordosis in the male hamster results from a deficiency in early steroids during critical periods of development. The reluctance to do so may arise from the fact that there clearly has been an amount of steroid present which is sufficient for the differentiation of normal male genitals. Accordingly, the bisexual behavior of the male hamster seems puzzling. Nevertheless, the experimental data indirectly support the hypothesis of a bisexuality originating from conditions of early steroid deficiency or early insensitivity to steroids present. In three laboratories currently studying the problem, the administration of additional steroids, either testosterone or estrogens, to the newborn male hamster has been associated with a decrease in bisexuality, as indicated by inhibition or suppression of lordosis responses in adulthood (Eaton, 1970; Swanson and Crossley, 1971; Paup *et al.,* 1972). The effect is not unique to the male hamster. Even though the expression of lordosis in the male rat is much more limited than in the male hamster, it can be reduced still further by the administration of steroid during the neonatal period (Hendricks, 1972).

Not all endocrine manipulations during early development act to suppress the development of lordosis and thereby decrease bisexuality of the genetic male. It is a well-documented principle that removal of the testes at a specific time in early development results in the development of lordosis in the genetic male without loss or diminution of mounting behavior. Thus deprivation of testicular steroids at a specific time can produce bisexual males. This effect of castration during the period of psychosexual differentiation does not produce a complete bisexuality, since such males fail to display all of the male sexual repertoire and intromission and ejaculatory responses are absent or deficient in adulthood. The effect of castration at this early developmental stage is reversible by treatment with exogenous steroids, provided that the replacement therapy is begun at once and not postponed to a later developmental stage.

Both the kind and amount of steroid hormone present in the developing male influence the degree of bisexuality that develops. When the testes are removed from the genetic male rat at birth and injections of relatively low amounts of androstenedione are given as replacement therapy, it is possible to produce an individual that displays lordosis readily in adulthood and that also displays mounting, intromission, and ejaculatory behavior (Goldfoot et al., 1969). If the amount of androstenedione injected during the neonatal period is increased, then the adult animal will not display lordosis (or will do so only to a very limited extent), and it will display the complete male sexual behavior repertoire (Stern, 1969). In other words, high concentrations of this steroid result in very limited bisexuality. Quite probably the same effects on development of bisexuality can be produced by quantitative variation of the amount of other steroid hormones such as estradiol or testosterone. Usually, however, these other steroids are so potent in suppressing lordosis that it is difficult in practical terms to find a dosage which both achieves full masculine development and fails to suppress lordosis.

The fact that males behave bisexually, even though they possess completely differentiated genitals, is not today as perplexing to behavioral endocrinologists as it was 10 years ago. The reason for the lack of concordance lies in the differing hormonal requirements for behavioral and genital systems. Despite the reliance of both systems on exposure to relevant steroids at a particular time in early development, it is possible that the two systems require different steroids and possibly different amounts of steroids at differing critical periods.

The possibility is not without experimental support. Estradiol benzoate administered at early stages of development is effective in reducing the expression of lordosis in both genetic males and females in adulthood, but the same steroid cannot cause normal phallic differentiation (Levine and Mullins, 1964; Paup et al., 1972). Two studies (Goldfoot et al., 1969; Stern, 1969) demonstrate that androstenedione in low to moderate amounts can induce good phallic differentiation in male rats castrated at birth but fails to suppress the expression of lordosis unless it is administered in high doses. Androstenedione, according to Stern (1969), has comparable effects in the genetic female and castrated male, the primary difference being that the phallus is not as differentiated in the female as in the male at birth.

The experimental induction of hermaphroditism is certainly germane to the issue of differing hormonal requirements for phallic and behavioral systems. Two examples from older studies provide evidence in support of this view. In the female guinea pig, Goy et al. (1964) showed that it was possible to separate effects on phallic differentiation and suppression of lordosis by varying the time in development when high concentrations of exogenous testosterone propionate were present. More recently, in the female rat, Whalen and Luttge (1971) and, in the guinea pig, Goldfoot and van der Werff ten Bosch (1975) showed that dihydrotestosterone propionate induced moderately good phallic differentiation without greatly impairing the expression of lordosis when the animals were mature.

PARAMETERS OF HORMONAL ACTION
DURING EARLY STAGES OF DEVELOPMENT

An extensive literature has developed which indicates that the major parameters of hormonal influence on development of mounting and lordosis are (1) concentration of hormonal substance in peripheral blood or at neural sites of action, (2) temporal aspects of early hormonal stimulation, and (3) chemical nature of the hormone at the site of action. Most of these variables have been identified by physiological experiments modifying the behavior of genetic females by treatment with exogenous steroids, or by modifying behavior of genetic males by castration with or without replacement procedures. Early in the work with experimental modification of the sexual behavior of the female guinea pig, the crucial nature of some of these variables was suggested (Goy et al., 1964). In those studies, it was found that high concentrations of exogenously administered testosterone propionate could at one stage of prenatal development inhibit the ultimate expression of lordosis without augmenting mounting behavior. At quite another stage of prenatal development, the same hormone in the same concentration augmented mounting behavior and had little or no effect on the later display of lordosis.

The empirical demonstration that mounting and lordosis behaviors are susceptible to hormonal influences at specific and different times in early development has suggested to some the existence of "critical periods." The concept of critical periods may not be relevant to hormonal effects on behavior in the sense that the necessity for the relevant hormone is limited to a very brief period in development and that the hormone is ineffective at any other time. We prefer the concept of a "period of maximal sensitivity" to hormonal influences, probably preceded and followed by periods of lesser sensitivity. During the periods of lesser sensitivity, the behaviors of mounting and lordosis can still be influenced by the relevant steroids, provided that high dosages are used. This view is consistent with Ward's view as well as with other data (Goy *et al.*, 1964) showing a relatively prolonged period even in the rat when steroids act to partially reduce the expression of lordosis in the adult. As is well known, however, even the periods of lesser sensitivity come to an end, and during other periods of development the hormones appear to be incapable of producing the same type of effect.

The overall picture that is emerging from laboratory studies of males from lower mammalian species can be summarized succinctly as follows: the degree of bisexuality expressed by the genetic male can be accounted for either by the amount and kind of steroid hormones present during a critical stage of psychosexual differentiation which occurs early in development or by a relative insensitivity to the steroid during this period. When the biologically active concentration of relevant steroids is high (either naturally or artificially elevated by injection), the adult male will display either very limited or no bisexuality as measured by the limited expression of lordosis and the frequent expression of mounting behavior. When the concentration of relevant steroids is low or the target tissues are insensitive, then the adult will be extensively bisexual; lordosis or presenting behavior will be frequently displayed and readily elicited and mounting can under appropriate circumstances also be displayed at frequencies judged to be normal for males of that species.

The hypothesis of tissue insensitivity seems to be a better model for the genetic male rhesus than a hypothesis based on absolute levels of steroid. Males from this higher primate species are highly bisexual, but their bisexuality seems not to be greatly influenced by the characteristics of the steroid environment of fetal life. Male rhesus monkeys, for example, display high levels of presenting behavior, even after extensive *in utero* exposure to testosterone or dihydrotestosterone. It is therefore possible that developing neural tissues destined to mediate presenting are insensitive to the steroids, and that in fact this may be a heritable trait developed under evolutionary selective pressures to protect the behavior from being suppressed. Wickler (1973) has suggested, for example, that male primates benefit from the ability to display presenting postures since it is used as a social *bonding* mechanism, important to the survival of the group.

Further evidence consistent with this hypothesis comes from studies of androgenization in the rhesus female. As in other species, androgen can masculinize the genitals and the behavior of a chromosomal female; but, unlike the case in other species, it cannot repress female responses, physiological or psychological. Thus a chromosomal rhesus female who is heavily androgenized both prenatally and in adult life nevertheless continues to menstruate (through the penis), to ovulate, and to exhibit feminine presenting behavior along with masculine mounting behavior. The presence of a gene which protects the relevant female tissues from androgen effects seems to us the most plausible explanation, and future experiments in this field should be focused in that direction.

RHESUS MONKEY AS A MODEL FOR OTHER
HUMAN SEXUAL BEHAVIORS[4]

In addition to serving as a model for bisexuality in the human species, the rhesus monkey may also in certain respects serve as a model for studies of human homosexuality, bachelorhood, and impotence.

Homosexuality

There can be little question that homosexuality in the behavioral sense requires a high degree of bisexuality at the neuroendocrinological level, and that this prerequisite exists in both the rhesus and the human species.

There is a process known to students of field behavior as *peripheralization of the male* which occurs in every troop that has been studied. Rhesus troops (like baboon troops) contain a group of males that form the apex of the central hierarchy — the leader males or alphas. Surrounding them are the females. All males not admitted to the central group are gradually forced to live on the edges of the troop; that is, they are "peripheralized." In a sense, they exist in a prison without walls, and homosexual behavior occurs very frequently in these peripheralized males.

The question of peripheralization in human societies, and of the effects of peripheralization on sexual behavior and life style, is one which should receive very great attention in connection with the etiology of homosexuality. Indeed, the process of alienation from fathers and peers among some feminine-behaving boys described earlier by Dr. Green is quite reminiscent of this process of peripheralization in rhesus troops. The differences between the sexuality characteristics of peripheralized male monkeys and homosexual human males may not be great. What is different is that the former are described and identified in terms of their behavior and geographic relationship to the leader males, whereas the latter are described in terms of their attitudinal relations to and feelings of alienation from so-called normal human males.

Bachelorhood

In addition to the males peripheralized to the edges of the rhesus troop, some males are further peripheralized to a state called *solitarization*. These monkeys become in effect hermits or bachelors.

In human societies, for some reason, bachelorhood is no longer considered a psychosexual disease (although someone once called celibacy the ultimate perversion). Bachelorhood, indeed, is no longer talked about as a problem. Very

[4] Dr. Goy made the following points during his oral presentation.

little concern or attention is paid to the sexual requirements or sexual role of the human bachelor. If such studies come back into style, the solitarized rhesus male should prove an excellent model for study, both in terms of the dynamics which lead to solitarization and in terms of the sexual consequences of solitarization.

As might be expected, the solitarized rhesus male is primarily autoerotic, but the frequencies and modes of autoerotic expression in these animals, as in human bachelors, have not to date been studied.

Impotence

Studies with nonhuman primates may provide some justification for the clinicians' views on human male impotence and its psychogenic origins and susceptibility to psychotherapeutic procedures. A very extensive study that Charles Phoenix, Adrien Slob, John Czaja, Kim Wallen, and R. W. Goy conducted on the sexual behavior of adult male rhesus monkeys illustrates the point. The study dealt with effects of castration (Phoenix *et al.*, 1973), but the data from the noncastrated controls are the focus of this discussion.

Nine feral males served as subjects. Approximately once each week each male was paired for a 10-min stand with one of eight different females. The procedure was repeated until every male had been paired eight times with each of the eight females, i.e., until each male had been paired a total of 64 times. The eight female partners were not allowed to undergo ovarian cycles. Instead, they were all ovariectomized and their endocrine condition was made equal by treating each with 13 daily injections of 10 μg estradiol benzoate before pairing with the males. A female was paired with a given male only at 2-month intervals, and all tests were not completed before 16–18 months had elapsed. This means that the overall performance shown was not based on tests given all on one day when either the male or female of the pair might be showing unusual kinds of behavior.

The numbers within the cells of Table I represent the frequency of ejaculation by each male with each female. Considered as a whole, they provide information on a characteristic of primate sexuality that Phoenix (1973) has called "compatibility" of the pair.

The development of the concept of compatibility is a judicious step in the analysis of nonhuman primate sexual behavior. The significance of these results in Table I cannot be revealed without the use of such a concept. For example, if we assume that there is no material difference between scores of 8 or 7 (since anyone can have a headache once in a while), then the differences among males cannot be accounted for by saying that different males have different maxima of sexual output. Every one of the nine males achieved seven or eight ejaculations out of his eight tests with one female or more. The differences among males, therefore, are not to be found in the maximum that can be achieved, but rather

Table I. Number of Occasions on Which Each Male Ejaculated When Paired Eight Different Times with Each Female[a]

Male No.	Female No.								Percent of total tests on which male ejaculated
	472	2220	2214	2224	2227	2229	2221	2203	
30	8	6	8	6	7	8	8	8	92.2
58	4	8	2	7	7	8	8	8	81.2
00	2	6	5	7	7	8	8	8	79.7
35	7	3	6	7	7	4	7	6	73.4
27	4	3	6	2	8	7	6	6	65.6
91	0	8	1	7	5	4	7	7	60.9
73	1	1	2	5	6	8	8	7	59.4
50	1	0	7	1	4	6	6	6	48.4
56	0	0	4	0	7	7	3	6	42.2
Percent of total tests that female received ejaculation	37.5	48.6	56.9	58.3	80.6	83.3	84.7	86.1	

[a]Data for these 576 tests were taken in part (564 tests) from Phoenix *et al.* (1973) and in part (12 tests) from Phoenix (1973).

in the availability of partners that permit, encourage, or stimulate the male to display his maximum expression. In short, the males do not differ in the attainable maximum, but pairs do, and the differences among pairs can be conceptualized along some dimension such as sexual compatibility.

What has been said above for the male is equally true for the female. Despite the fact that female 2224 seems to have had a headache more often than most (since no male ejaculated with her on all eight tests), none of the females showed consistent rejection and frigidity.

The conclusion is compelling, for nonhuman primates such as the rhesus anyway, that profound, consistent, and universal sexual apathy is not a frequently encountered condition. Moreover, a reasonable parallel exists between the performances of some of these rhesus pairs and the performances that clinical workers encounter among human beings. For example, male 50 and female 472 would neither be judged nor judge themselves as having a highly satisfactory sexual relationship. At least it would seem that only very lax standards would allow an incidence of one ejaculation out of every eight opportunities to be regarded as highly satisfactory. If male 50 were to "divorce" or separate from female 472 and establish a "legal consort bond" with female 2220, the sexual relationship might not improve and in fact might be considered worse. Our test results show that when this arrangement was made for him, he never achieved ejaculation with 2220. Interestingly, male 56 has a nearly exact replication of this history with these two female partners. My guess is that any clinician who heard a history from a human male that was limited to the experiences of these rhesus males with these rhesus females would begin to suspect the patient had serious libidinal problems. It would not be unreasonable under such circumstances to label the case file "male impotence" or "female impotence" depending on the sex of the patient. The correctness of this diagnosis depends on the implications the label carries as well as on what it connotes for causation of the condition. If the label of "impotence" carries only the connotation of partner incompatibility, then in the rhesus cases being discussed it would be correct since males 50 and 56 ejaculated on seven out of eight possible occasions when paired, respectively, with female 2214 or females 2227 and 2229.

The matrix presented in Table I could be thought of as a "couples test," and the percentages given in the right-hand column show that the males achieve different performance levels on this test. Similarly, percentages in the bottom row of Table I show that females achieve different performance levels. These individual differences among males and among females are influenced not only by occasional "impotence" but by another factor as well. In the case of either the male or the female, the marginal percentages are not the result of a failure to achieve the maximum but they are influenced greatly by how consistent the scores for a given individual are. This may be the best operational definition that an experimentalist can give for level of libido or sexual drive, but the differences can be conceptualized alternatively as differences in "sexual finickiness." Pairs may differ in compatibility, but individuals differ in the finickiness of their sexuality. If compatibility is defined, as Phoenix has suggested, as directly related to the proportion of tests with a given partner on which ejaculation occurs, then "finickiness" might be measured by some variable such as the dispersion or variance of scores with different partners. Thus it is possible to redefine impotence of one sort in terms of pair compatibility, and libido can be redefined as sexual finickiness. We do not know or understand the factors that establish or maintain either pair compatibility or finickiness, but both are certainly important variables in primate sexuality. Some workers with nonhuman primates in the field have pointed to factors such as social status that may be related to compatibility, but it is doubtful that this variable influences outcomes in the testing situation used in this study. The whole question needs to be held in abeyance for the moment.

A final word of caution must be injected into this discussion. The concepts of sexual compatibility and sexual finickiness are neither adequate nor very useful at extremes. For example, if a male were to be tested eight times with each of eight females and failed to ejaculate on any test, then no decision could be reached regarding whether finickiness or compatibility was the factor responsible for the overall poor performance. In such a case a concept of "profound impotence" might as well be resorted to as a diagnostic category. This is a different

sort of impotence from the sort described above for two male rhesus and from
the sort most commonly encountered in clinical practice. As with all diagnostic
terms in this field, the label means nothing with respect to etiology, and in its
best usage merely connotes a kind of performance under reasonably standard
conditions.

Group Discussion

Dr. Fordney-Settlage asked about finickiness in the female monkey. Dr.
Goy replied that he knew of no studies in the natural habitat. In the laboratory,
the female rhesus seems to be very finicky during the luteal phase of her cycle
and early in the follicular phase; thereafter, she defies all of the evolutionary
guidelines governing genetic selection and copulates with any willing male.

Dr. Goy was asked whether the zero scores in the table might represent
female rejection rather than male failure of potency. He replied that there was
no way to tell.

Dr. Lipman-Blumen asked Dr. Goy about the interval between exposures
for the female. He replied that the female was usually tested with five males on
one day, then not tested further for 3 weeks or so. Five males per day, however,
is far below the rhesus female's capacity; in nontest situations she may copulate
with 20 males a day with no trouble at all.

Dr. Rose reported that his monkey observations were performed on large
monkey troops in large compounds rather than on pairs or trios, and that perhaps for this reason his interpretations are quite different. The difference might
be illustrated by supposing that the human handshake, if brief, remains a form
of greeting behavior, but when prolonged leads to orgasm. This would not con-
vert greeting behavior into sexual behavior. Similarly in his monkey troops,
much of what Dr. Goy scores as sexual behavior can be scored as dominant or
submissive behavior. Presentation behavior, for example, occurs very often in
both males and females as part of a chain of behavior in which a dominant animal approaches a submissive animal who responds by averting his gaze and then
presenting. The dominant animal may then mount. If intromission and coitus do
not follow, however, it is very difficult to say that this is sexual rather than
dominant-submissive behavior. Other observations confirm the importance of
distinguishing dominant-submissive presenting or mounting from sexual presenting or mounting. If this distinction is made, there may be less bisexuality observed among monkeys.

Dr. Goy replied that he agreed with Dr. Rose's observations, but thought
the distinction between sexual and dominance behavior too sharp. Perhaps in
human sexuality, too, there is a substantial dominance factor. Accepting the dominance function of presenting and mounting does not detract from their sexual
nature. They are both dominance-related and sexual. Sexuality, in the rhesus as

in the human, can be exhibited in a context of dominant behavior, submissive behavior, greeting behavior, and so on, as well as in the context of sexual gratification. Moreover, at least some homosexual behavior between male rhesus monkeys is unquestionably sexual in the strictest sense. Thus a 45-sec film sequence of two peripheralized rhesus males shows not only mounting but also anal intromission and thrusting to ejaculation on the part of the mounting monkey while the presenting monkey is simultaneously masturbating to orgasm. This sequence can hardly be confused with dominance or greeting behavior.

Dr. Rose raised the question of whether homosexual behavior in the male rhesus might not be wholly facultative — that is, engaged in *faute de mieux*. He knew of no data showing that a rhesus male with free access to a female in estrus would by preference mount and engage in anal coitus with another male.

Dr. Goy pointed out that the position he tried to develop was that all homosexual behavior, in monkeys and men, is "facultative." The difference between monkeys and men on the one hand and lower mammals on the other is that the neuroendocrine basis for bisexuality exists to a lesser degree in lower mammals and hence in those forms there is less likelihood of "facultative" facilitation of a homosexual pattern. In any case, as Dr. Rose pointed out, it is all for the best.

Dr. Rose called attention to recent studies indicating that social experience can alter androgen levels. Thus following a fight which he loses the androgen level of a male may fall and remain low until he engages in another fight which he wins. Dr. Goy noted that in some of his experiments males displayed presenting behavior at times when their testosterone level was high, as determined by assay.

Dr. Rose commented that, in his opinion, monkeys should be studied in a natural setting; caged monkeys are crazy. Dr. Goy replied that "crazy" is perhaps an exaggeration, and the term "legally insane" is more accurate.

Dr. Cole asked about the sexual activity of physically handicapped monkeys. Dr. Goy cited the example of a monkey female, a member of a macaque troop in a natural setting, with a neurological disease which paralyzed her hindquarters. While not a highly prized sexual object, she had twice delivered infants since her paralysis. This was true despite the fact that males were unable to copulate with her in the usual way. Instead, it was necessary for the male to grasp her hindquarters, raise them up, and then move them forward and backward on the intromitted penis.

Dr. Rubinstein speculated that infrahuman species might be arranged in a hierarchy from lower to higher, with an increasingly variegated range of behavior available at the higher levels. This might be true in particular of capacity for varied forms of sexual behavior. When one reaches the human species, however, another factor enters — social constraints on variations in sexual behavior. Thus the paradox would emerge that the species inherently capable of the most varied sexual behavior might in fact exhibit the fewest variations. Further, a comparison of the human species with other species high on the scale might throw light on the extent of this repression by social constraints.

Attention was called to recent studies in which pregnant rats were subjected to restraints and other stresses during pregnancy. Male offspring of these rats showed deficiencies in male behavior. Whether or not these deficiencies are traceable to prenatal hormone levels, it was suggested, the finding that pregnancy conditions can affect the masculinity of offspring is a clue which should be followed up.

Dr. Green asked about mother-son incest in rhesus troops in natural settings. Dr. Goy replied that it occurs, but only rarely. Since 80% of a troop's males leave their native troop at adolescence, only 20% have an opportunity for mother-son incest. In rhesus as in human societies, Dr. Goy added, the mother is socially dominant in relation to her sons, and sexual behavior rarely occurs where the female is the dominant member of a pair. Thus no magical or biological incest taboo is needed to explain the observed rarity of mother-son incest.

Research in Homosexuality: Back to the Drawing Board[1]

Alan P. Bell, Ph.D.[2]

Before addressing ourselves to the question "Where do we go from here?" with respect to research in the area of homosexuality, we would do well to take stock of where we have been and where we are. In this regard, I cannot think of a more comprehensive statement than what is to be found in the preface to Weinberg's and my annotated bibliography of homosexuality. In our summary of the 1265 items which were included in that volume, we pointed out that

> discussions of homosexuality have consisted primarily of speculations prompted by theoretical models or statements whose constructs have not been tested in any systematic manner.... Studies designed to test these assumptions [about the nature of homosexual development] have been few, while those which have been conducted have usually included small, biased samples as well as measurements which have been subjectively derived. Little attention has been given to the wide range of homosexual orientation and adjustment; most have viewed homosexuality-heterosexuality as a simple dichotomy ... most of their subjects have been those who eschew their homosexual orientation and whose functioning in other areas of their lives has been marginal. Usually there has been no attempt to determine the relationship of etiological factors to subsequent behavior and adjustment. In addition, the focus of these researchers and commentators all too often has precluded any reference to those processes — both sociological and psychological — which maintain the homosexual's career. The homosexual is most commonly viewed as an inheritor of certain dispositions which were crystallized in the past and which account inevitably for all of his future behavior ... a consideration of his behaviors as goal-directed is generally disregarded....
> [There has been no attempt] to investigate systematically differences in the developmental histories, in the life styles and adjustments, in the sexual attitudes and behaviors, and in the wide range of self- and other-experiencing, which may prove to be functions of such matters as age, sex, education, and race, which transcend whatever important features exist in the homosexual community.
> ... a most apparent characteristic of that research [pertaining to homosexuality] has been its lack of coordination ... this tendency to work apart from others has resulted in

---

[1] This paper was presented at the conference, "Sex Research: Future Directions," held at the State University of New York at Stony Brook, Stony Brook, New York, June 5-9, 1974.

[2] Senior Research Psychologist, Institute for Sex Research, Indiana University, Bloomington, Indiana 47401.

an incredible waste of time, talent, and human resources within the research enterprise. Isolated findings remain unrelated. Important areas of inquiry remain unacknowledged. The fund of knowledge and insight which might serve as a foundation for new and important advances in research is not increased. Differences between the findings of various researchers which may stem from different instrumentation and subject populations remain uninvestigated. . . . most researchers are loath to replicate their findings. . . . [Thus] conclusions are often stated which, given the unreliability of instrumentation and the methods of sample selection, are unwarranted. . . . [Such] conclusions — uncertainly derived and nowhere replicated — often determine the nature and direction of further inquiry whose parameters of investigation may involve no more than the compounding of error. Scientific inquiry remains such in name only if it is not accompanied by opportunities to measure its phenomena in different times and circumstances. (Weinberg and Bell, 1972, pp. xi-xiii)

This view of the matter will be reflected in the discussion which follows. It is based on thoughts which have emerged slowly over the course of the more than several years that I have been involved in this area of research, on my perusal of the recommendations of the NIMH Task Force on Homosexuality (1972), as well as on conversations with my several colleagues at the Institute. What I would like us to consider is the need for a coordinated, interdisciplinary effort which promises a more valid and comprehensive appraisal of what it means to be homosexual: sexually, socially, and psychologically. I wish that I were in a position to report that, on the basis of my own and others' research in this area, we were now in a position to proceed with a mop-up operation in this area of sex research. Nothing could be farther from the truth. Rather, my general sense of the matter is that perhaps we have reached the point where "going back to the drawing board" makes the most sense.

If it were up to me, I would declare a moratorium on the usual conduct of research in homosexuality, at least over the next several years. The veritable flood of investigation in this area which has recently appeared would be reduced to a trickle, replaced by a monumental attention to methodological issues pertaining to the research enterprise. Those invited back to the drawing board — clinicians and behavioral scientists of every kind, physiologists, criminologists, etc. — would be asked to address themselves to a variety of questions and to become engaged in a variety of tasks. First, they would be asked to take stock of whatever efforts have been made to assess the homosexual phenomenon, of the methodologies employed, and of researchers' results and conclusions. After the population of inquiries had been ascertained and delineated, an effort would then be made to explore the feasibility of reaching anything like a consensus about how and in what direction future investigations might be pursued. This would involve, for example, specifying the parameters of homosexual experience, determining the ways in which they might be most properly measured, and considering the extent to which comparable methodologies might be employed by persons possessing similar as well as idiosyncratic research interests. The underlying questions to which I would have one and all address themselves include the following: To what extent do our theoretical orientations and research strategies prevent us from becoming fully cognizant of the diversity of homosexual experience? What is there about who we are and what we think we know that restricts unduly the scope of our inquiry? In what ways and at what point can a view of the forest promote a view of the trees, and *vice versa?*

Before addressing myself to these questions, I would suggest that we shall never know much more about the nature of homosexual experience in either broad or highly specific ways until we delineate the sexual aspects of that experience much more precisely. No comprehensive assessment of homosexuality can ever be made, no in-depth studies will ever be especially informative, until we have investigated the number and nature of the sexual parameters associated with homosexuality *per se,* until we are in a position to create a viable typology. Certainly studies which view homosexuality vs. heterosexuality as a simple dichotomy, which lump homosexuals together simply on the basis of their sexual object choice, should be eschewed.

Homosexuals differ with respect to the degree to which they are exclusively homosexual in their sexual arousal and behavior. These differences are reflected in the nature of their past and present sexual fantasies and dreams, in their past and present ratings on the so-called Kinsey scale, in the degree to which they have ever been able to respond sexually to a person of the opposite sex, and in the emotional contexts in which such arousal is most likely to occur. Given such differences, a series of methodological studies relating to the degree of a person's homosexual orientation is very much in order. What are the relationships between the several aspects of homosexual orientation which I have just enumerated? On what basis does a person consider himself or herself a 6 or a 5 or a 4 on the Kinsey scale? The reasons for such self-ratings may be much more informative than the self-ratings themselves. When a person is asked to rate himself or herself on the Kinsey scale, is he or she comparing himself or herself with someone else? Is the person comparing his present standing to where he was at another period of his life? To what extent is the person's self-rating related to his ideal self? To a need to avoid ambiguity? To the social circumstances in which the report is given? Will the person rate himself differently with an interviewer who is perceived as "straight" than with a person who is thought to represent a gay alliance group? What is the relationship between an individual's self-report and such objectively determined criteria as certain measurable physiological responses made under experimental conditions? Although self-ratings of this kind are used in innumerable studies, few if any attempts have been made to understand the variety of factors which might prompt such reports. This is a good example of the kind of thing I have in mind when I speak of "going back to the drawing board."

Other ways in which homosexuals differ "sexually" include (1) the number of their partners, (2) the physical characteristics they seek in their partners, (3) the degree of intimacy obtained in a given relationship, (4) the emotional meaning of their sexual partnerships, (5) the locales in which prospective partners are sought and in which sexual contact takes place, (6) the frequency with which sexual contacts are pursued and found, (7) the extensiveness of their sexual repertoires as well as their preferred sexual techniques, (8) the degree to which their sexual impulses are ego-alien, (9) the extent and nature of their sexual problems, (10) the level of their sexual interest and arousal, and (11) the degree to which they are overt or covert. Each of these parameters of the homosexual experience deserves extensive investigation, as do their relationships to each other. It would be my hope that as a result of a number of different studies involving many different samples, and employing relevant factor-analysis strategies, an appraisal instrument could be devised — perhaps akin to the MMPI — which would make it possible for researchers to compare homosexuals on the basis of their sexual profiles and to specify more precisely a person's place in the homosexual universe. A comparable instrument could and should be devised for heterosexuals as well, an instrument which would be of great value to clinicians as well as to researchers investigating the panoply of sexual behavior. I am convinced that differences between people with regard to one or another of these sexual parameters will be far more revealing and informative than whether a given experience occurs in a homosexual or heterosexual context.

Only after we have really homed in on the diverse nature of homosexual sexual experience and devised an instrument which taps that diversity in a reliable fashion should we proceed to the question of what additional variables should be used in an effort to better understand that experience. The number and kinds of variables we wish to include will, of course, depend in large measure on the nature of our theoretical perspectives.

Some persons have claimed that up until the past 5 years homosexuality has been construed largely in sexual terms, that insufficient attention has been given to the nonsexual aspects of the homosexual experience. I do not believe this to be the case at all. The fact is, most of the literature pertaining to homosexuality has consisted of data obtained by psychoanalytically oriented clinicians who conceptualize homosexuality almost entirely with reference to oedipal issues, who view homosexual behavior as a defensive maneuver associated with very early experiences of parents as objects of erotic stimulation and of identification. They have tended to view unconscious conflict emerging from the nature of one's earliest relationships with parents as the chief, if not exclusive, motivating factor and to consider every other aspect of homosexual experience as simply a reiteration of early infantile concerns. It is not that they have failed to attend to the nonsexual aspects of homosexuality. In fact, the opposite is the case. From their perspective, motivations for homosexuality have a dreamlike quality, reflecting an assortment of intrapsychic needs having little to do with explicitly sexual satisfactions. Their theoretical orientation — which specifies a narrow set of underlying motivations for various sexual behaviors as well as the means by which these motivations can be uncovered — has led to a preoccupation with etiology, and for at least two reasons: (1) the present circumstances of any behavior are thought to be overshadowed by the past, and to be fraught with goals having little to do with a person's present external environment; and (2) the nature of their clinical enterprise includes a profound attention to and understanding of infancy and early childhood.

Less orthodox Freudians and other neoanalytic theorists and clinicians, impatient with such a narrow view of psychosexual development which does not coincide with their own data-gathering expeditions, have enlarged the number of parameters thought to be related to the origin and maintenance of homosexual behavior. For example, Hatterer (1970) lists 27 different variables pertaining to parental relationships which could have a bearing on homosexual development, most of which are not exactly related to Freud's oedipal terminology. He lists 50 other variables, some having to do with a search for an appropriate gender identity, others with sibling relationships, and still others with various cultural and environmental factors having little to do with the familial milieu.

In the Institute's so-called San Francisco Study I have borrowed heavily from the psychodynamic view of sexual development, but my sociologist colleagues have also made sure that we are in a position to investigate the data with reference to conditioning and labeling theory as well. Here again I am reminded of the importance of the need for interdisciplinary research.

I am very much interested in the etiology of homosexuality — or, more properly, the origins of diverse patterns of sexual arousal and behavior — and refuse to be put off by those who are convinced that attention to the past is neither interesting nor rewarding or by gay radicals who question the motivations of those who seek answers to the nature of homosexual development. We are beginning to make some progress in this area, especially in the growing awareness that, as Hatterer and others suggest, there are probably multiple routes into a given sexual orientation. I believe that Saghir and Robins' (1973) finding that homosexuals and heterosexuals, both male and female, could be differentiated on the basis of romantic fantasies and cognitional rehearsals occurring during childhood and early adolescence provides important clues about the nature of homosexual and heterosexual development. In contrast to the recommendations of the NIMH Task Force — which placed problems of etiology and determinants at the bottom of their list of recommended research projects — I would strongly urge additional, innovative work in this area. It might well include (1) one or more longitudinal studies (akin to Richard Green's work on cross-sex experience in childhood) in which large numbers of prepubertal boys and girls are followed through young adulthood, (2) a more systematic study of the nature of retrospective reporting, including the use of hypnosis with adult samples, (3) a greater use of multisource data (parents, siblings, peers), (4) studies designed to determine the extent to which different methodologies (questionnaires, structured interviews, unstructured interviews, extended clinical contacts, direct observations, etc.) tend to produce different pictures of childhood and adolescence. It is very clear that any studies of the origins of sexual

patterns should include a highly detailed inquiry, probably involving more attention to covert processes than to overt behavior. In addition to subjects' chronological ages, it would be important to relate a broad range of social and sexual
experience to the ages of puberty, of first homosexual and/or heterosexual contacts, of subjects' first romantic attachments, and of their first cognitional rehearsals. Ideally, whatever attempts are made to delineate the various developmental routes of sociosexual development should include an effort to explore
their relationship to those aspects of adult sexual functioning which have already
been mentioned.

After enumerating the possible determinants of male homosexuality, Hatterer (1970) concludes that, while an individual's family can make him vulnera-
ble to homosexuality, "family influence does not necessarily explain why homosexual habits become permanent. The cause of that must be sought among hun-
dreds of other variables concerning the man himself and the world he lives in"
(pp. 42-43). This important point, stated in different ways in more recent investigations of homosexuality, implies the need to conceive of sexual patterns as
not simply the result of past influences but as goal-directed behavior involving a
broad range of reinforcements in the present which are not entirely related to
residues of the past and which involve more than the fulfillment of sexual needs.
Elsewhere (1974) I have referred to differences among homosexuals with regard
to what is figure and what is ground at any given point in their sociosexual development. This notion is very much related to the present plea that the social and
psychological meanings which homosexuality now has for each individual be
thoroughly explored in an attempt to construct a truly comprehensive typology.
One person's homosexuality may amount chiefly to a political statement.
Another's may represent a quest for intimacy. Still another's may be indistin-
guishable from a preoccupation with occupational needs, interests, and rewards.
Researchers must be encouraged to listen with a "third ear" to their subjects'
reports and with an understanding that the explanation of subjects' behaviors
need not be limited to a delineation of intrapsychic issues.

This larger view of the matter makes it imperative that homosexual patterns be related to the cultural milieu in which they occur. In this regard, perhaps it would make sense for prospective investigators to become totally im-
mersed in various gay worlds and in the lives of its different memberships. Up
until now there has been a tendency for researchers to view from too great a
distance "the man himself and the world he lives in," to be more adept at data
analysis than at acquainting themselves firsthand with the incredible variety of
homosexual experience and coming up with truly meaningful variables. I myself
feel terribly remote from those whom I am in the process of studying, writing
from lists reporting significant differences between various samples of individuals
I have never met and who have lived out their lives in a social setting I have never
known. Because of this, and because of a variety of personal and theoretical
biases which keep me uninformed, I am convinced that we have failed to tap

important dimensions of the homosexual experience. More contact with our subject populations would probably have enhanced my own self-awareness as well as my awareness of the diversity of homosexual experience; our study would thus have been at once more objective and comprehensive. I would certainly hope that those sociologists, anthropologists, and social psychologists who assemble at the "drawing board" are not like certain criminologists who have never seen the inside of a prison or certain urban sociologists who spend more of their time in small towns like Bloomington, Indiana. I would hope that their deliberations could lead to the specification of those broad parameters of social experience (and of how they might be measured) which should be included in any comprehensive investigation of homosexuality or in connection with studies characterized by more depth and less scope.

Carrying this a step further, I see the need for two distinct but related emphases in future research. On the one hand, there is the need for studies which focus on the forest, whose scope is broad. I have in mind large-scale survey research projects, touching all of the bases in a relatively superficial way, but with an instrumentation which is both reliable and valid. In this connection, too much of our work seems to involve the "reinvention of the wheel," too many home-brewed instrumentations which may appeal to a given researcher's ego, but which hardly contribute to an advancement of the research enterprise. I can envision the establishment of an "item bank" — a pool of questions designed to tap the social, psychological, and sexual dimensions of the homosexual experience — from which individual researchers could draw for at least certain areas of their inquiry, as well as a "data bank" where it would be possible to replicate various constellations of findings obtained from different samples. The kind of survey research I have in mind would involve samples in which the diversity of homosexual experience was represented and would always include a certain amount of attention to a variety of methodological questions, such as how to make certain samples more accessible for research, the effectiveness of various recruitment procedures, the effects of interviewer characteristics or respondent-interviewer interaction on the data, or the basis on which individuals report a given feeling state. Efforts along these lines might also include an investigation of the feasibility of obtaining homosexual samples by random probability techniques through the use of some sort of screening device.

One real importance of survey research is that it provides a grid in which to locate an individual or a smaller group of individuals, an opportunity to describe in detail a smaller segment of the population with reference to a host of normative data. And this I see as the other important research thrust of the future. We need to go deeper in a given area, to refine and elaborate on whatever areas of investigation are suggested by studies characterized by a wider scope of inquiry. In addition to the investigation of special groups of homosexuals, I see the need for in-depth case studies of individuals which reveal the predictive power (or the lack of it) of large-scale findings for a given individual. The attempt to "flesh out" normative data in this way could prove to be a humbling yet highly informative experience. Many people, including psychoanalysts and other clinicians, could be involved in such an activity, which, in turn, would help to promote a much-needed dialogue between persons who, at this point, tend to distrust each other's data.

I suppose that all of my previous remarks could be pretty well summarized as follows: Before we shall ever get a very clear picture of the influence of certain independent variables on homosexuality — whether they be present or past social or psychological circumstances, or whatever efforts are employed to modify various aspects of homosexual experience — the dependent or criterion variable must be ascertained much more precisely than it is at the present time.

The Institute's San Francisco Study on which I am currently at work is primarily a study of the forest rather than of individual trees. Our sample of 1000 was drawn from a pool of some 5000 homosexuals whom we contacted in the San Francisco Bay area, male and female, black and white, aged 17 through 65, recruited in a wide variety of ways. We sampled the lists of the major homophile organizations, for example. We recruited at every one of San Francisco's gay bars, at gay theaters, private parties, in public toilets, at churches, among motorcycle gangs, and so on. We spread the word of our survey throughout the community — by radio, television, the newspapers, and stickers in washrooms reading "KINSEY IS HERE," together with a phone number. To maximize the likelihood and quality of usable responses, we employed male and female interviewers, black and white, straight and gay, to interview corresponding sections of the homosexual and control populations. The selection of San Francisco — a relatively "good" scene for homosexuals — made it possible to reach homosexuals who might have remained inaccessible elsewhere. Thus, whatever the shortcomings of our sample, it is substantially larger and more variegated than in most prior studies. We also have a heterosexual control group, of more than 500, female and male, black and white, recruited from door to door.

The interview questionnaire we used is 175 pages long and takes on the average about 4 hr to administer, including open-ended as well as checklist questions. It is, of course, wholly inadequate for extracting from respondents the full richness of their experience, but it is certainly fuller than most of its predecessors. It is expected that our final report, one volume on homosexual development and another on homosexuals, will total a thousand pages. Thus again, whatever its shortcomings, our San Francisco Study should at the very least be useful in planning both more comprehensive and more intensive future studies.

Group Discussion

Dr. Bell was asked about matching his homosexual subjects with heterosexuals in the control group. If unmarried homosexual males are compared only with unmarried heterosexual males, for example, one atypical group is being matched with another. Dr. Bell replied that he will be able to compare homosexual males with both unmarried and married heterosexual males.

Dr. Bell was asked about the validity of the answers given to emotionally laden questions. He replied that he had applied for a supplementary grant to explore this crucial question but had not received it. Hence all he could say in his report would be that "Homosexuals, with respect to this variable, tend to report the following to a greater or lesser degree than heterosexuals, or report in a different way."

Dr. Bell was asked whether his study would throw light on the relative prevalence of depression, anxiety, or other symptoms of psychological dysfunction in homosexuals as compared with heterosexuals. He replied that his questionnaire covered depression, anxiety, psychosomatic illnesses, and many related parameters.

Dr. Bell called attention to the fact that his study is not based on homosexual-heterosexual dichotomy. While most of his homosexuals are 6's on the Kinsey scale (exclusively homosexual) and most of his controls are 0's (exclusively heterosexual), there are enough in categories 1 through 5 to demonstrate the continuum, including various degrees of bisexuality. The study will report, for example, that black homosexuals and female homosexuals are less likely to describe themselves as exclusively homosexual than white male homosexuals.

Dr. Bell was asked what motives, other than intellectual curiosity, motivated the study. He replied that he hoped it would have a practical impact on the oppressive situation homosexuals face, and on public attitudes toward homosexuality.

He was also asked how he reconciled the need for confidentiality with the need to avoid including the same individual twice in the sample. He replied that cards bearing the names or pseudonyms and addresses of respondents accompanied the questionnaire returns from San Francisco to Bloomington and were then destroyed. Thus it was theoretically possible for an individual to be included under two pseudonyms. Destroying the cards of course made a follow-up study impossible.

Dr. Bell was asked why he selected a retrospective strategy in the light of the well-known defects of retrospective studies. He replied that, despite its deficiencies, it is the only strategy available for surveying a large homosexual population. Moreover, the study is only in part concerned with retrospective data; it is also in considerable part concerned with the current life patterns of homosexuals and heterosexuals.

Dr. Green regretted that homosexuals picked up in the random sampling procedures which generated the heterosexual control group were not studied and compared with the homosexuals recruited in other ways. Dr. Bell also expressed regret that this had not been done.

Dr. Gebhard called attention to another Kinsey Institute study headed by Albert D. Klassen and Eugene Levitt, based on a national random sample of

3000 — piggybacked on a study of the same sample by the National Opinion Research Corporation (NORC). The primary purpose of the study was to determine public attitudes and biases toward particular types of sexual behavior. It seemed likely that attitude would be highly correlated with whether a respondent had had a particular type of sexual experience, so additional questions were asked concerning sexual experiences, including homosexual experiences. Such a study can supply nationwide random-sample answers; but, unlike the San Francisco Study, it can supply only crude yes-no, ever-never kinds of answers.

Dr. Bell was asked what course he would take if his data showed homosexuals to be highly neurotic or psychotic individuals, deeply damaged in various ways. During the discussion which followed, it was noted that this is a particular case of a very general problem: what to do with data which are potentially harmful socially or to a social group. Dr. James suggested that the truth is rarely harmful; proponents of a point of view who are basing their programs on false data are benefited rather than harmed when better data are supplied them. In the case of the homosexual study, it was pointed out that at the very least the data would show that all homosexuals are not alike, and that there is a continuum connecting extreme homosexuals with extreme heterosexuals — two findings which in themselves should prove beneficial by destroying stereotypes.

Dr. Rubinstein felt that researchers can become unduly concerned that their findings might prove socially harmful. In the first place, a very long road separates research findings from social action; research findings are rarely decisive in the social arena. In the second place, each reader brings to the findings his own bias and interprets them in his own way. The researcher, even though he may try, can rarely dictate the social response which his findings will evoke. Indeed, an objective statement of findings may prove far more effective in producing social change than a report deliberately drafted with that goal in mind.

The value was suggested of using comparable questions in various surveys so that the results can be compared. Dr. Bell agreed, and said that many of the questions in his survey paralleled those asked in earlier surveys; hence results can indeed be compared.

Dr. Green raised a question concerning a prediction by Dr. Bell that his homosexual sample will show greater childhood stress and more disturbed parent-child relations than his heterosexual sample. Suppose, he said, that this is also true of others who exhibit types of behavior labeled deviant in our culture. In that case, childhood stress and disturbed parent-child relations might explain subsequent deviance, but some other factor might determine whether this deviance took the form of homosexuality or another form. Dr. Lipman-Blumen pointed out that such factors have been found in the childhood of juvenile delinquents, schizophrenics, and so on, but that such studies are increasingly being challenged and are not in good repute. It was also pointed out that the fact that an outcome was associated with childhood stress or disturbed parent-child relations did not *ipso facto* warrant the conclusion that the outcome was "bad" or "sick." It is conceivable that childhood stress and disturbed parent-child relations might lead to "good," "healthy" outcomes.

Dr. Bell was asked whether, out of the infinity of questions which might be asked homosexuals, there was a theoretical framework which guided the selection of the questions selected for his study. Dr. Bell replied that the theoretical framework was primarily psychoanalytic, but with labeling theory and other points of view also influential in construction of the questionnaire.

Dr. Gebhard called attention to the fact that the San Francisco Study was funded by the National Institute of Mental Health; other sources of funding had been approached but had lacked the courage to fund a large-scale objective study of homosexuality.

Changing Sex Roles in American Culture: Future Directions for Research[1]

Jean Lipman-Blumen, Ph.D.[2]

PRESENT STATUS OF SEX ROLE RESEARCH

Sex role research, in the past, has been multidimensional, but misleading, and often mythical in its foundations. The study of sex roles traditionally has been in large measure the bailiwick of sociologists. They have tended to define sex roles generally in terms of differentiation — first, within the family (Parsons, 1942; Parsons and Bales, 1955) and, second, within the economy (Oppenheimer, 1968; Epstein, 1971). Within theoretical frameworks provided largely by the functionalist school of sociology, this approach has viewed sex role as one form of social role in which there are reciprocal behaviors and attitudes, governed by articulated norms and rewarded in terms of differential contributions and values. This perspective leans heavily on role theory (Biddle and Thomas, 1966) and tends to focus on such concepts as socialization (Brim, 1968; Goslin, 1969), role conflict (Gross et al., 1958), identification (Lynn, 1969), role models and role set (Merton, 1957), role differentiation (Eisenstadt, 1971; Holter, 1970), and role dedifferentiation (Lipman-Blumen, 1973). The socialization literature, subsumed under this perspective, is a major body of literature in its own right, with social learning theory (Mischel, 1966) and cognitive mode theory (Kohlberg, 1966) representing at least two major thrusts designed to explain how children learn and assume "appropriate" sex roles.

Another dimension of sex role research has been developed by anthropologists who have tried to partial out the effects of culture on sex roles by examining sex role within a cross-cultural matrix (Mead, 1935).

More recently, social scientists of various disciplines have been examining alternative modes of sex role patterning, including communes (Kanter, 1972),

[1] This paper was presented at the conference, "Sex Research: Future Directions," held at the State University of New York at Stony Brook, Stony Brook, New York, June 5-9, 1974.
[2] Office of Research, National Institute of Education, Washington, D.C. 20208.

kibbutzim (Spiro, 1956; Talmon, 1965; Mednick, in press), female communities, and homosexual and lesbian unions and parenting.

Another subdomain of sex role research historically developed through research on the family. Some of the earliest work on the family (particularly studies based on the notion that women needed expert advice on how to carry out their roles as wives and mothers) originated within the ranks of home economists. Both the functionalist sociology theorists and the home economists, for different reasons, wove questionable assumptions into this body of literature, which is being reviewed and reconceptualized in the wake of recent criticism.

Women as a minority group represent a more recent third tributary of the sex role literature. This perspective examines the dynamics of minority group membership, including discrimination, prejudice, marginality, and assimilation (Hacker, 1951; Myrdal, 1944; Andreas, 1971). This literature leans heavily on theoretical perspectives developed in the study of various racial, ethnic, and religious minorities.

A fourth and recent perspective understands sex roles in terms of the "politics of caste" (Hochschild, 1973), in which power, interest, and resource differentials are seen as the basis of exploitation of the lower caste (females) by the higher caste (males) (Collins, 1971). It is within this context that sometimes appropriate and sometimes inappropriate analogies are drawn between women and blacks, and women and slaves. Only a small group of social scientists have brought the theoretical constructs from the study of stratification (Acker, 1973) and mobility to bear on the study of sex roles, despite the abundance of empirical descriptive material documenting women's subordinate status within the educational, occupational, and economic structures of our society.

Another research tributary has been the study of sex differences, dominated mainly by psychologists and focused on emotive, cognitive, and physical traits (Maccoby, 1966; Garai and Scheinfeld, 1968). This body of literature placed strong emphasis on an assumed "genetic imperative" underlying sex differentiation. In addition, major generalizations about female-male differences have been made on the basis of laboratory experiments, many involving game situations or laboratory tasks removed from "real life" activities for which they purportedly served as analogues.

Social scientists tend to focus on "significant differences" (even when they are substantively trivial) and rarely report lack of differences. As a result, similarities of the sexes, as well as the meaning of the overlap of distributions on various dimensions, rarely are discussed.

The women's movement has stimulated serious debate about sex roles and the assumptions underlying the very experiments and instruments by which sex differences allegedly have been documented. As a result, this large corpus of work recently has undergone reevaluation (Maccoby and Jacklin, 1974a,b).

A closely related perspective deals with gender roles and gender identification, an area developed largely by psychologists, psychiatrists, and some sociologists, who have seen hormonal levels, physiology, and family patterns as correlates of gender identification (Stoller, 1968; Green, 1974; Green and Money, 1969; Chafetz, 1974). This general subdomain also has drawn attention to the issues of transsexuality, homosexuality, and transvestism.

Another source of sex role discussion, often subjective and polemical, but nonetheless valuable, is the feminist literature, which has seen a renaissance in recent years. This literature, while not usually falling within the recognized canons of scientific inquiry, takes a critical view of scientific dicta developed within the previous perspectives and demands that "traditional" social and natural sciences respond to its critiques and insights (de Beauvoir, 1953; Friedan, 1963; Millet, 1970; Greer, 1971; Firestone, 1970).

It is largely from this source — feminist criticism — that the assumptions of "scientific" sex role research have been brought into question, thereby sensitizing the scientific community to its "trained incapacity," and its own blind spots. The community of scientists historically has not taken kindly to questioning of its sacred dicta, even when the questioning came from *within* the scientific community. Semmelweise, Lister, and Freud are scientists who experienced the rebuff of the scientific community, because their work represented either a departure from or a critique of "establishment" science. If the scientific community is reluctant to accept such criticism from within, it is not surprising that it is even less receptive to the questioning of its concepts and methods from outsiders, even when those outsiders represent the very population the scientific world is trying to explain. Therefore, it is a welcome sign that this source of criticism originally from outside — and more and more from inside — the scientific community is having a noticeable impact on the direction and quality of research on sex roles.

WHITHER SEX ROLE RESEARCH?

But where does this leave us? And where is sex role research heading? Or, perhaps more properly, where *should* it be going? I would like to propose a set of research strategies and substantive foci that might enhance both the quality and the direction of research in this area. The first eight strategies are designed to build a systematic knowledge base, while the ninth strategy attempts to implement the knowledge developed within the first eight.

Strategy 1: Critical Synthesis

Critical synthesis of the descriptive and analytical data on the multiple reflections of sex differences, sex role differentiation, women as a minority, stratification, mobility and discrimination, gender roles, and gender identification.

While much work has been done in these different areas, the burgeoning mass of research is difficult, if not impossible, to synthesize and assimilate. It is necessary to invest the effort to bring it together in some coherent form, under meaningful categories of analysis. Such synthesis provides necessary guides to where our knowledge is dependable, where unexplored areas remain, and where bridges must be built to greater understanding.

Strategy 2: Process Analysis

Analysis of the processes which underlie and contribute to various forms of sex differences, sex role partitioning, the formation and perpetuation of women as a minority or deviant group, as well as sex role stratification.

This strategy involves development and testing of analytical models of these processes. Without articulation of these processes, it is not feasible to think in terms of developing intervention modes.

What work has been done on processes is often contradictory and remains to be empirically tested, refined, and integrated. The conditions under which a given process leads to a "traditional" differentiated result and the conditions necessary for variations in and departures from traditional sex roles are important lacunae in our knowledge.

This emphasis on process cuts across the existing perspectives within sex role research. We need to understand the processes by which sex differences, which are not demonstrable during early childhood, emerge in adolescence and later. We also need to explicate those processes by which early differences become accentuated with successive developmental stages.

The processual strategy is applicable to sex role research, particularly in attempting to delineate changes in sex roles as both females and males move through succeeding stages in the life cycle. It is important to explicate how, why, and if sex role changes are more easily tolerated and integrated by women or men in different life stages.

Process analysis is equally valid in examining the minority group status of women and problems of stratification and mobility. It is a tool for understanding the development of a "career" as a minority or deviant group member. It is a means of looking at downward, as well as upward, mobility, with the built-in bonus of permitting understanding of the dynamics of nonmobility and role stagnation.

Strategy 3: Social System/Personal
System Integration

Integration of social system and personal system variables, including documentation and analysis of the sociopsychological, economic, political, racial-

ethnic, historical, chronological, and legal factors that are linked to sex and sex role differences in attitudes and behaviors, including patterns of leadership and cognitive styles, achievements, aptitudes, and aspirations.

While considerable work has been devoted to sex-linked differences in cognitive style, achievement motivation, and sexual permissiveness, these personal system variables often are analyzed in a vacuum. Little attempt is made to understand the complex relationships between individual or group differences and factors emanating from the larger social system within which these individuals and groups develop. Oftentimes researchers act as if these differences (or similarities, as we now are beginning to recognize) can be understood either in total laboratory isolation or solely within the context of the family.

Much research, particularly on sex roles *per se,* has suffered from the fact that sex roles often are studied and seen as inexorably embedded in the family. Perhaps one reason social scientists see little change in sex roles is because they look at sex roles within the setting that is least likely to generate or tolerate change in both sex and generational roles — the family. The study of sex roles must be conducted within nonfamily settings (e.g., nonmarried living together arrangements, the occupational world) that reflect the impact of other social system variables unmediated by the structure of the family.

While we ask for sex roles to be considered outside the structural parameters of families, we also would recommend a more realistic and meaningful appraisal of sex roles *within the family.* Much family research has been conducted on the basis of interviews, questionnaires, laboratory games, and pencil-and-paper tests. Considerably less family research has been conducted within the physical empirical world of the family. We suspect that long-term participant observation studies of families will yield new insights and understanding of the realities of sex roles within families. One piece of evidence that supports such a view is the work recently reported (Maccoby and Jacklin, 1974b) in which husband-wife teams were compared to "stranger" teams instructed to play the roles of husband and wife. It was the "stranger" teams who fell into the stereotypical roles we understand as "husband" and "wife"; the real-life marital pairs acted in freer, less stereotyped ways.

When we take sex role studies outside the family structure and begin to examine them in other settings — in the world of work, politics, and the economy — we begin to enter the arena of what is probably forbidden knowledge. Without the protection of the normative setting of the family, we can begin to question the meaning and impact of sex roles for all the other roles that are superimposed on sex roles: for colleague roles, for friendship roles, for noncolleague work roles, for political roles, for consumer roles, etc.

Within this general strategy, there is a serious need for comparative research on the male sex role. Despite the lip-service paid to the need for investigations of the male role, until recently, little substantive work has been done, with some obvious exceptions (Aldous, 1969; Dahlstrom, 1967; Turner, 1970; Benson, 1972; Grønseth, 1971/1972; Tresemer and Pleck, 1972; Weiss, 1973; Bart, in progress). As the focus and parameters of women's roles continue to expand, research on the male role becomes more imperative in order to understand the reciprocal changes for men. Considerable attention has been focused on the tensions and conflicts involved in the wife-mother (sometimes worker) role; however, a relative paucity of work exists on comparable role strain and ambiguity experienced by men. The study of role conflict and strain in men's roles has been limited primarily to male occupational roles, with further delimitation to the top segment of the occupational spectrum (roles highly valued in our society).

While we suggested under Strategy 2 the need for examining the development of careers among deviant individuals and groups, here we would recommend that social scientists address the question of similarities among deviant, disabled, and disadvantaged groups. Many of the concepts that we apply to the deviant and disabled that deal with their ability or inability to act independently, their special gifts, their need to be protected from society, society's need to be protected from them, as well as the need to sequester these "exceptional" groups from the mainstream of society, should be reexamined in terms of their applicability to women within this and other cultures. It might be equally useful to compare the social legislation growing out of these conceptions of the criminal, the aged, the emotionally ill, and the physically handicapped to social legislation affecting women.

Strategy 4: Cross-Cultural and Interdisciplinary Analysis

Cross-cultural and interdisciplinary comparisons of sex role patterns, including sex role learning and expectations in different cultural settings that allow us to separate out those features that persist across all cultures and to understand the implications of this universality.

The rationale for a cross-cultural research effort in the area of sex roles rests on the need to examine in detail the "genetic imperative" assumption that underlines much sex role differentiation analysis. Cross-cultural methodology would serve nicely to separate out universal from situational or more parochial patterns in sex role configurations. Cross-cultural research can offer us important insights into those roles considered "deviant" or marginal within our own culture which are highly valued in other cultures.

In addition to laying the genetic imperative assumption to rest, cross-cultural research has other distinctive advantages. It fosters an understanding of both *how* roles may be configured differently and *why* they are. It illuminates the mechanisms underlying alternative role patterns and provides an understanding of processes. Major cross-national studies of early childhood, adolescence, youth, and maturity should be undertaken which use comparable design and instrumentation.

In cross-cultural as well as in other modes of research, the difficult but rewarding interdisciplinary approach is worth pursuing. Cross-cultural studies are particularly amenable to interdisciplinary efforts, where historical, psychological, economic, sociological and political science, as well as physical and medical perspectives, add significant breadth and depth to our understanding of sex roles.

Cross-cultural comparisons among societies in different stages of economic and political development represent a crucial source of knowledge. They enlarge our perception of how sex roles are articulated with the economic and political systems, and how they change in consonance or dissonance with shifts in the macrostructure. The implications of older marriages, more childless marriages, fewer children, and greater longevity of women should be explored within a cross-cultural perspective to see how other societies handle these demographic trends.

Strategy 5: New Theories of the Middle Range

Development of theories of the middle range that delineate a limited segment of society or social action (i.e., sex roles) but attempt to break through the confines of previous research and political limitations.

Many years ago, Merton (1949) explicated the value of developing theories of the middle range, efforts to explain a limited phenomenon as opposed to grand theories which attempt to explain all of society within one consistent and coherent framework. Within this strategy, new, radical, and perhaps not so radical, middle-range theories should be developed and opened for discussion if only to permit new currents and maybe provocation within a sometimes conceptually tired field. This may be the most difficult strategy of all, because of the obvious impediments to dispensing with our usual cultural and intellectual, not to mention political and emotional, blinders.

Here I shall sketch briefly an example of such a middle-range theory, one that I call a Homosocial View of Sex Roles (Lipman-Blumen, in progress). This theoretical framework suggests that males have a predisposition to be interested in, excited by, or stimulated by other males. This predisposition to be interested in other males then is fostered and reinforced by a social structure characterized by a stratification system in which males are more highly valued individuals, occupy more highly valued roles, and have almost exclusive access to the major spectra of resources. Contrary to the frequent accusation that men have turned women into sex objects, women are forced by the structural situation into the role of sex object in order to distract men into entering and developing hetero-sexual relationships.

Men can turn to other men for the satisfaction of most of their needs: intellectual, physical, emotional, social, and sexual stimulation, as well as power, status, and the broad nexus of occupational, political, and economic needs. The one element that men can not derive from other men is the legitimation of their masculinity in terms of fatherhood. Potential paternity is the major re-source (not simply sexuality, as is usually suggested) that women can offer to men to gain entry into their world, to share their status and power, as well as their beds.

Men, as well as women, perceive men as the controllers of various types of resources, and this is simply a hard-headed recognition of the actual reward structure within society. Until quite recently, women and men both identified with other men, hardly surprising in view of research that indicates that individ-uals tend to identify with those persons whom they perceive as controlling re-sources. Neither women nor men found it meaningful or helpful to identify with women who lacked most major means of enhancing life conditions.

Sociologists often conceptualize relationships as exchange systems. Prior to the feminist movement, the basis of exchange between males and females was decidedly asymmetrical. Men could offer into the bargain their power and status, based on education, occupation, and income (all higher than women's). They could add to this their intellectualism, competitiveness, leadership, and aggressiv-ity. Finally, they could contribute potential maternity (reciprocally the legitima-tion of a woman's femininity), but not necessarily sexuality. Women, on the other hand, had only two major resources: potential paternity and sexuality. This uneven array of resources systematically made men more interesting to women, women less interesting and useful to other women, and women fairly often unnecessary and/or burdensome to men.

The women's movement, by pressing for women's greater access to educa-tional, occupational, economic, political, and legal sources of power, has made a first attempt to redress this imbalance. In so doing, women's roles are being somewhat more highly valued (perhaps only in certain quarters). Women who are able to help other women, to dispense power and status and other forms of aid and support, suddenly are seen anew by both other women and men. Women thus become more important to one another, and their powers for exchange with men are greatly strengthened. We have the foundations of a parallel struc-tured homosocial orientation for women. In essence, we are witnessing a social crisis, a condition which has been suggested as a sufficient cause of role change or role dedifferentiation (Lipman-Blumen, 1973). The stratification system is be-ing significantly altered by the strains of the crisis, which requires a reallocation of resources between the sexes.

This middle-range framework of a homosocial view of sex roles is explic-itly distinct from Tiger's (1970) bonding theory. Further, it is not intended as a euphemistic way of suggesting that men are predisposed to homosexuality [We have borrowed Gagnon and Simon's term "homosocial" (1967) specifically to underscore this distinction.]

But what is the evidence for this homosocial view of sex roles? The evidence comes from several sources, including early childhood development studies, patterns of adolescent and adult interaction of the two sexes, as well as animal studies (for those who are willing to extrapolate from monkeys to men). Summarizing from a broad range of sex differences research shows that boys more often attempt to dominate other boys, boys more often pick other boys as victims (despite equally nonreinforcing response patterns from both male and female victims), boys' activity level increases when they are in the company of other boys, boys' competitiveness increases when in all-male groups, boys' aggression is stimulated by the presence of other boys, boys' reading problems are ameliorated when they are treated in male settings as opposed to female or individualized settings. Attempts by males to dominate are more often directed toward each other or toward adults (although less often). Dominance attempts toward girls are infrequent. In one study of 2½-year-old children in a nursery school setting (Halverson and Waldrop, 1973), there was no difference by sex in activity when the children were playing alone. When playing with same-sex peers, boys' activity level significantly increased, but this was not true for girls.

Another study suggests that 18-month-old boys hug, stroke, and kiss each other. Eighteen-month-old girls do not exhibit similar behavior. By 3 years of age, the stroking, hugging, and kissing by boys has been transformed into rough-and-tumble play.

Maccoby and Jacklin interpret this stroking-kissing-hugging behavior as "preaggressive play," an earlier necessary developmental stage. While the behavior is preaggressive in the temporal sense, we would argue that it is preaggressive for somewhat different reasons than those they propose. It is preaggressive behavior because the only way in which our present society can tolerate males touching one another is in an aggressive, competitive manner. Transforming natural affectionate behavior into aggressive rough-and-tumble behavior is one way of protecting valued males from falling into the (wasteful) trap of homosexuality. (The taboos against lesbianism are not nearly as stringent, partly because female roles are seen as potentially dispensable, or at least less valuable, in our society.)

In adulthood, "normal" men reportedly dream twice as often about men as about women. (Women dream equally about the two sexes.) In both male and female dreams about aggression, the aggressor more frequently is pictured as male (Hall and Van Der Castle, 1966).

More evidence is found in suggestive monkey studies by Mitchell and Brandt (1970). They claim that male and female infant monkeys have equal activity levels when they are caged alone or only with their mothers. However, when they are placed in cages where they are separated from another mother and infant by a glass panel enabling them to see the other pair, the activity level of male infant pairs is higher and involves play-imitating and threats toward one another.

In conclusion, it is obvious that boys are stimulated by other boys. In encounters with each other their activity level increases, and they play roughly, attempt to dominate one another, and fight. Women do not show the same sort of activity patterns, nor do cross-sex pairs.

In light of these findings, it is not difficult to begin to understand women's absence from the world of finance, politics, and ideas, *not* as a deliberate hostile conspiracy against women, but as the manifestation of a self-sufficient, homosocial male society. This is perhaps a too-lengthy example of new middle-range approaches, stemming from existing research and suggesting a new organizing hypothesis that brings coherence to data not previously necessarily recognized as interrelated.

Strategy 6: Mapping Sex Roles onto Sexual Behavior (and Vice Versa)

Integration of studies of sexuality and sexual behavior with research on sex roles (and other social roles) to better understand the interdependence of the two levels of behavior.

Since individuals express their sexuality within the parameters of defined sex roles and other social roles, it is only logical that we should approach the study of sexual behavior within the natural context of sex roles and *vice versa.* At the present time, an unfortunate disciplinary dichotomy exists whereby, in most cases, sociologists tend to study sex roles isolated from sexuality and sexual behavior, and psychologists, psychiatrists, and endocrinologists investigate sexual behavior apart from sex roles.

An integration of these two substantive fields of inquiry would illuminate the processes involved in changes in sex roles as one moves through the life cycle, along with adaptive (or maladaptive) changes in sexual behavior. This approach would foster understanding of what happens to an individual's sexual needs as she or he moves from one developmental life stage to another. Are different sexual modalities, or combinations, more appropriate at different stages of the life cycle? As an individual moves from the role of spouse to that of widow(er), homosexual unions or autoeroticism may represent more viable choices, through either personal preference or lack of alternatives.

This strategy would permit us to understand how sex roles, as well as the entire range of social roles, constrain or shape our sexuality. Occupational, political, community, and other social roles set limits or offer opportunities for sexuality. Social scientists, with few exceptions (Cuber and Harroff, 1966), have ignored this area, as if it were the apple of forbidden knowledge.

Strategy 7: Specification of Conditions
for Changing Sex Roles

Enumeration and specification of conditions under which sex roles are most likely to change, providing perspectives on how the entire range of social roles changes.

This strategy is an important tool for the eventual redesign of sex roles and other social roles. In focusing on the intrinsic interest in sex roles, we have neglected to emphasize the importance of sex roles as a useful paradigm for all social roles. Thus an understanding of how sex roles may undergo change allows us to plan for broad-spectrum transitions in all roles, without social chaos. Analysis of necessary or sufficient conditions for role change enables us to meet changes when they occur without social and personal dislocation.

Strategy 8: Multidimensional Methodologies

Development of appropriate, complex, multidimensional methodologies to meet the intrinsically complex problems of sex role research.

Our growing recognition of the complexity of sex roles makes it imperative that we develop the appropriate methodologies to deal with this area. This strategy would involve designing more cross-cultural and interdisciplinary research, as noted above. In addition, it calls for the integration of numerous approaches: longitudinal and cross-sectional studies; qualitative and quantitative methods; experimental, ethological, and ethnological techniques; interviews, questionnaires, and observation. This strategy also calls for more attention to problems of sampling and control groups. Within such a strategy, we should attempt to develop large-scale national data bases on changes in sexual behavior, comparable to the census data we collect on demographic variables. This would enable us to relate demographic changes to sex role and sexual behavior trends.

Strategy 9: Intervention and Planned Change Research

Development of techniques, procedures, and social policy recommendations aimed at changing the traditional sex role differentiations that currently limit both females and males in their life situations.

This strategy is necessary in order to implement the findings developed within the previous strategies described above. Indeed, the other strategies may be viewed as preludes to intervention and planned change.

Implementation of findings through resocialization techniques as well as legislative intervention represents an important research initiative. Otherwise, we are forced to raise the question "Knowledge for what?" This strategy allows us the opportunity to develop training techniques and procedures for changing behavior and eventually attitudes on the basis of systematic knowledge developed within the eight prior strategies. Resocialization techniques can be developed and tested, and longitudinal studies should be designed to evaluate the efficacy and permanence of the behavioral and attitudinal changes thus effected. Social policy recommendations should be developed on the basis of the knowledge that emerges from the previous strategies. Following enactment of legislation bearing on sex roles (i.e., Title IX), longitudinal studies should be conducted to measure the impact of these legal changes.

Existing social policy should be reviewed to assess its impact on sex roles, and the implications of sex role research and sex role changes for social policy should be evaluated. Historical examples of the relationship between legislation and sex role behavior, such as in Japan after World War II, should be carried out within this strategy.

Summation: Clearly this program is ambitious. Certain segments are more difficult to implement and accomplish than others. The reconceptualization of an area so long delineated in human practice is a formidable task; the testing and implementation of this reconceptualization will perhaps be even more arduous and lengthy.

But the *need* for restructuring sex roles is even greater and creates an urgency about this task which we ought to delay no longer.

Group Discussion

Dr. Lipman-Blumen was asked to outline a project, using only finite resources, to test the hypothesis that the homosocial behavior of males — their preference for the company of other males rather than of women — results from the fact that males control the reward structure. She replied that she is hardly at the design stage for such a project, but she could conceive of several preliminary approaches. For example, a study of the social consequences of abortion clinics run by females offers a readily available small-scale test. At such a clinic, many women see for the first time other women controlling resources and controlling the reward structure; the result is the emergence of a homosocial attitude among the women. The same occurs in the ambience of women psychiatrists and of other women in a position to allocate resources and control rewards. This could be studied in detail. Years ago, women physicians had difficulty developing a female patient load, for women patients as well as men saw men physicians as more powerful. All this is changing as women increasingly see one another as actually in control of resources.

Dr. James noted that this is not just a hypothesis but a current phenomenon. As women take over control of gynecological clinics, other women flock to them for their obstetrical and gynecological care — to such an extent that male-controlled clinics must now seek desperately for female obstetrician-gynecologists in order to hold their patients. A female gynecologist or psychiatrist may have a hard time attracting patients in a male-controlled setting; when the control of the resource is in female hands, the reluctance of women to patronize women physicians disappears.

Dr. Lipman-Blumen suggested another simple test model in which male and female participants in a game or task situation are given equal control over resources on some occasions, and control is lodged exclusively with the males or exclusively with the females on other occasions. What will be the differential effects on female and male behavior?

Dr. Gebhard noted that you can find anything in anthropology if you look for it, including cultures where males exhibit behavior which in our culture is labeled typically female. While we are almost running out of preliterate societies, studies could still be made in a few places of the interrelations between control over resource allocation and sex role behavior.

Dr. Goodman called attention to another view of homosocial behavior among males. The initial sex role of both males and females, according to this view, is female, as a result of the close association of both with their mothers. Male children must later learn to be male, and they learn this from one another. Dr. James replied that at the cognitive level the lesson children learn may be the reverse. Observing the inferior status of the mother in the family constellation, they may turn away from that role.

Mr. Brecher suggested that before seeking the origins of homosocial behavior among heterosexual males it would be useful to describe and measure that behavior. A study which followed 100 normal heterosexual American males for 1 week might reveal that they spend an enormous proportion of their time reacting with other males in a broad variety of ways — as distinct from the relatively few ways they interact with females during a much smaller proportion of their time. Further, such a study might identify some deviant males who are heterosocial as well as heterosexual. And, finally, it might turn out that some homosexual males are heterosocial, preferring the companionship of women except for sexual encounters. Dr. Lipman-Blumen suggested the concept of a homosocial-heterosocial continuum similar to the homosexual-heterosexual continuum.

Dr. Lipman-Blumen was asked if she wanted to repress aggressivity as such. She answered no; she was instead concerned to remedy a situation in which some characteristics, such as aggressivity, were assigned exclusively to males and others, such as passivity, exclusively to females. McClelland and others studying "need for achievement" (N.Ach.) in the human species have been surprised to find that women respond so differently from men in test situations where N.Ach. is measured. This difference, she suggested, may result from the fact that females from an early age are programmed to accept *vicarious* achievement roles, or enabling roles (Lipman-Blumen, 1975). The occupational roles labeled female in our culture are the enabling facilitating roles — roles offering only vicarious satisfaction of the need for achievement. We keep females locked into these vicarious roles by demeaning their sexual identity if they try to escape. Just as there is no reason why women should be limited to vicarious achievement, so there is no reason why men should not be comfortable in roles which offer them a sense of vicarious achievement.

When we examine what is actually happening, Dr. Lipman-Blumen continued, we see that men are even more tightly locked into stereotyped roles than are women. A woman who enters a male profession, for example, is given negative brownie points for feminine sexuality, but her effectiveness on the job is nevertheless recognized. We say, "She's not much of a woman, but she's certainly an achiever." When a male enters a facilitating or ancillary occupational role such as nursing, however, we give him a double negative, demeaning both his aggressivity and his sexual masculinity. Moreover, when a male does enter what is a traditionally female occupation, society makes every effort to rescue him from his supposedly ignominious position; he is very rapidly promoted into a supervisory role. This is true of the male nurse and the male teacher. While this purportedly rescues men from the double negative, it may also deprive them of the satisfactions they were seeking when they initially decided to become nurses or teachers.

Dr. Rose pointed out that in monkey troops in a natural setting males are in general dominant over females — but this does not mean that females are powerless. On the contrary, females in the alpha group are dominant over the nonalpha males. They have a major role in shaping the rise to dominance of the male offspring — a much greater role than males have. The alpha females also control some of the resources of the group. If they get together, they can prevent a male from achieving a dominant position. This female control over male dominance in turn determines access to resources.

Comprehensive Sex Research Centers: Design and Operation for Effective Functioning[1]

Paul H. Gebhard, Ph.D.[2]

INTRODUCTION

To my way of thinking, a sex research center must consist of (1) two or more professionally trained individuals, preferably with advanced degrees, and (2) a support staff; the center must (3) have a separate legal identity or, if it is part of a host organization (e.g., university, hospital, clinic), be recognized by that organization as a separate (although not necessarily autonomous) unit, and (4) have the ability to devote a substantial amount of time to sex research. These four criteria are the bare minimum to warrant the title "sex research center." There are numerous instances of informal associations which verge on being sex research centers. For example, several scientists or clinicians in the same institution who share an interest in sex research and who may have independent research projects in that field are very close to constituting a research center. All they lack is a formally acknowledged sense of being a coordinated group with interdependencies and common goals. Most sex research centers have developed out of such a situation. Similarly there are existing organizations which need only to devote more time to research to qualify as sex research centers. This is true of various educational organizations such as SIECUS, the Centro Studi Educazione Sessuale, and therapeutic organizations such as sex clinics, marriage counseling groups, and Planned Parenthood. In order to distinguish between an educational or clinical organization which sponsors an occasional piece of research and a true sex research center, I submit that to merit the latter title at least one-quarter of the group's time and effort must be devoted to sex research and the group must consider such research to be one of its goals and not merely an adjunct activity.

[1] This paper was presented at the conference, "Sex Research: Future Directions," held at the State University of New York at Stony Brook, Stony Brook, New York, June 5-9, 1974.
[2] Institute for Sex Research, Indiana University, Bloomington, Indiana 47401.

The term "comprehensive" I should like to define as meaning broad in both the disciplines utilized and the research topics chosen. In brief, a comprehensive center is one which can launch a multidisciplinary attack on a diversity of sexual topics. This definition rules out many research centers, some because they are confined to one approach (e.g., psychoanalysis) or because they limit themselves in terms of topic (e.g., the Erickson Foundation, whose interests lie almost wholly in transexualism and transvestism).

STATUS OF ORGANIZED SEX RESEARCH

Human sex research at present is limited chiefly to the wealthier nations. It is an unaffordable luxury to developing countries which must devote their money and effort to providing a subsistence diet and at least minimal health services to their inhabitants. Consequently, vast areas of Africa, India, and Indonesia are not yet capable of supporting sex research except on forms of contraception.

However, even in nations which can afford sex research, the forces of prudery, conservatism, and inertia have combined to severely hamper it. To choose a dramatic example, Saudi Arabia could finance a huge sex research organization and scarcely notice the expenditure, but the chances of this occurring are less than those of a snowball in Saudi Arabia.

Antisexualism also explains the absence of sex research centers in Central and South America, where, however, there are indications of the beginnings of sex research — generally under the aegis of education and/or medicine. We know of no sex research organizations in China, but the restraint there seems to be political. This is no surprise: nearly all totalitarian states regard sexuality as an enemy or at least as an unfortunate necessity which must not be allowed to unduly distract the populace from more important undertakings.

In brief, in surveying the status of sex research of any magnitude one may justifiably look only at Europe, Japan, and North America.

In Scandinavia, only Denmark and Sweden are involved to any extent. Sweden boasts a score of professionals doing some work in the field and a number of organizations such as the Group for Family Research, the Public Committee on Sex Education, the Swedish Institute for Sexual Research, and the Swedish Institute for Sexual Information (RFSU). However, these organizations emphasize treatment and education; research is definitely in third place. Denmark has a dozen or more persons active in sexual studies, and Preben Hertoft is organizing an institute.

West Germany has at least 30 persons doing some research and an impressive number of organizations: the Institut für Sexualforschung at the University of Hamburg founded by Hans Giese and carried on by Eberhard Schorsch, Gunter Schmidt, and Volkmar Sigusch (Sigusch is now at the Abteilung für Medizinische Sexualwissenschaft at the University of Frankfurt); a Society for the Advancement of Social Scientific Sex Research (located in Dusseldorf); and the Deutsche Gesellschaft für Sexualforschung, which is an almost exact parallel of the Society for the Scientific Study of Sex in the United States. The Gesellschaft publishes a journal and there is another West German periodical devoted to sex — *Sexualmedizin.*

The lowland nations of Northwestern Europe are heavily involved both in lobbying for sexual tolerance and in research. Holland has the Netherlands Institute for Social and Sexual Research (NISSO) and there is a chair in sexology at the University of Amsterdam occupied by Conrad van Emde Boas. In Belgium there is the Center for Interdisciplinary Study of Sex and at the University of Louvain there is the Center d'Etude Familiales et Sexologiques.

France has few persons active in sex research, but they are making a valiant effort to launch sex research and a sex research organization via a July 1974 meeting in Paris.

England has a rather large number of interested persons, many of whom are engaged in individual research, but almost no centers or national organizations have emerged aside from some homophile groups. Efforts are currently under way, chiefly by John Bancroft, to remedy this situation, and Martin Cole has set up his Institute for Sex Education and Research.

Switzerland has the newly organized Center for Sex Research founded by Willy Pasini, Georges Abraham, and William Geisendorf at the medical school in Geneva. There is also a Center for Medico-Social Family Sex Education.

Italian sex research faces great difficulties, but, nevertheless, two organizations have emerged and several international meetings have been held. The organizations are the Centro Studi Educazione Sessuale and the Centro Italiano di Sessuologia. There is also one sex research journal, *Sessuologia,* edited by Romano Forleo.

Spain and Portugal are even less hospitable to sex research than Italy. In 1971 Nicholas Caparros was arrested and charged with creating a public scandal when he distributed questionnaires at the University of Madrid.

The East European communist bloc countries present a mixed picture and accurate information is hard to collect. There appears to be an embryonic group in Bulgaria, but none in Yugoslavia. I have no data for Roumania. A moderate amount of good sex research is being carried on in East Germany and in Poland, but it seems to be on an individual basis and I know of no institutes or other research organizations except for the newly formed Scientific Wheel of Polish Sexologists, headed by Boleslaw Popielski. Czechoslovakia, on the other hand, has a large and long-established Sexological Institute at Charles University in Prague. Founded by Joseph Hynie, it is now headed by Jan Raboch. The staff, numbering about eight, is primarily involved in clinical work, but many manage to find some time for research. This institute is in an unusually favorable position for research since all sexological cases in the country are referred there.

The USSR in recent years has inaugurated some sex research by a small number of individuals. However, the nation is establishing a large number of clinics to treat sexual problems and, in addition, has a Department of Sexual Pathology at the University of Moscow (headed by Georg Vassilchenko) and an institute at Tbilisi. It is anticipated that these clinics and other agencies will engender a large amount of sex research which will probably rival or surpass in quantity that of any other European country.

In Israel several persons are conducting sex research on an individual basis, but the emphasis is on sex education. A second International Symposium on Sex Education was held in 1974.

Japan has had a Japanese Society for Sex Research and a Life Psychology Society for some years, but these do not appear to be scientific; they are more for the lay public. In 1972, however, a truly professional organization was founded: the Japanese Association for Sex Education. The chairman is Tsuneo Uchida, formerly Minister of Health and Welfare, and the Executive Director is Shinichi Asayama, a professor emeritus known for an earlier sex survey of Japanese students. This association intends to do sex research as well as educate.

Turning finally to North America, there is little organized research in Canada, but there are a number of potentially important developments. Several faculty members at York University are researching; there is an Institute of Sexology in Montreal; and the University of Quebec has a lengthy curriculum devoted to sexology.

The situation in the United States is complex. Now that sex research has become almost respectable, and in some ways prestigious, numerous individuals have entered the field and organizations have proliferated at an increasing rate. It is almost impossible to obtain adequate information on some of these — for instance, I know the Council for the Study of Human Sexuality as only a post office box number in Nashville, Tennessee. Other organizations appear to be operated by only one or two individuals, e.g., the American Interpersonal Institute at Sarasota, Florida, and the Center for Marital and Sexual Studies, which Hartman and Fithian operate in Long Beach, California.

Other organizations are larger and better known, but ascertaining the amount of research they do is often difficult. It is not clear how much of the University of Minnesota research output is from the Program in Human Sexuality (a loose consortium of Minnesota faculty headed by Richard Chilgren) and how much represents effort outside that program. The same problem obtains with regard to Dr. Schumacher's Human Sexuality Center at Long Island Jewish–Hillside Medical Center, which in terms of numbers has a large professional staff. (Dr. Schumacher recently went to the University of Pittsburgh.) Basically, in these and similar cases, one faces the question of whether there is interdependence and some team effort or whether the organization is simply a nominal umbrella over a series of independent researchers.

It is pointless to differentiate sex research organizations on the basis of the degree of autonomy accorded the constituent members. It is better to identify research-productive groups regardless of their structure. There are at least seven such groups in the United States today which merit attention because of their productivity and/or potential:

1. The Johns Hopkins group, which grew out of the Gender Identity Clinic and the work of John Money.
2. The UCLA group consisting of Robert Stoller and Richard Green, with Green having recently departed for Stony Brook.
3. The Institute for Sex Research at Indiana University, founded by Alfred Kinsey.
4. The Center for the Study of Sex Education in Medicine located in Philadelphia, founded by Harold Lief.
5. The Reproductive Biology Research Foundation in St. Louis, established by William Masters and Virginia Johnson.

To this list of older groups one must add two new organizations:

6. The Human Sexuality Center at Long Island Jewish–Hillside Medical Center.
7. Our host, the Stony Brook group, founded by Stanley Yolles.

This list could expand dramatically and quickly since at several places, notably the University of Minnesota and the University of Washington, there are a number of proven researchers who could coalesce and form a strong group if someone provided the money and organizational effort.

NEED FOR COMPREHENSIVE SEX RESEARCH CENTERS

While the foregoing list of seven major research groups is impressive, one should not be misled into unfounded optimism as to the status of human sex research in the United States. First, the scope of a number of these groups is limited, for they focus on specific topics. Thus the Johns Hopkins and UCLA groups emphasize gender identity and role development; Lief's group confines itself to research on medical sex education; and the Reproductive Biology Research Foundation has been primarily concerned with physiological and therapeutic matters. The Institute for Sex Research seems the lone generalist or comprehensive center at the moment, but there is reason to expect that the Stony Brook people will through their diversity of interests become comprehensive, also. It is hard to prophesy about the Human Sexuality Center, but I suspect that its clinical orientation may confine its research to patient populations and problems.

In addition to these topical limitations, there is the crucial limitation of research time. The great majority, if not all, of the senior researchers of these organizations have substantial commitments to teaching, therapy, and administration. Thus a person known as a sex researcher may in actuality devote only a quarter or less of his or her time to research.

If sex research is to be effectively facilitated, it is obvious that, first, we need more researchers so that their fragments of research time will sum to an adequate total. Second, I believe that researchers will function more efficiently in a comprehensive multidisciplinary center. The field of human sexuality draws on so many disciplines that the lone researcher, competent only in his own field, is in grave danger of myopia and error. It is not sufficient simply to have all or most relevant disciplines on the same campus or in the same hospital; there must be some organizational structure binding people together with reciprocal obligations.

A comprehensive research center need not be generalistic incessantly — it can focus on specific projects, completing them serially. The important thing is to retain the generalist approach, the vital breadth of knowledge, interest, and view, even while concentrating on the current projects.

NEED FOR A SEX RESEARCH CONSORTIUM

Even if more comprehensive research centers are developed, there is need for still one other organizational accomplishment — the formulation of some consortium relationship between research groups and ultimately between individual investigators. At present, nearly every group or person labors with little or no idea of what others are doing. Our efforts at the Institute for Sex Research to develop an international directory of researchers with some identification of their current interest and work have shown us the degree of isolation we all suffer. As a result, we cannot profit from one another's accomplishments or failures, we engage in wasteful overlapping work, and we miss intellectual cross-fertilization.

Organizations such as the Deutsche Gesellschaft, SIECUS, the Czechoslovakian Sexological Society, and the Society for the Scientific Study of Sex do not meet this need. Moreover, I doubt that they could be modified to meet the need since they do not consist primarily of researchers. What is needed is some international organization formed for and by research persons expressly for communication and coordination. Fortunately, such an organization, the International Academy of Sex Research, has been formulated by Richard Green; its voice will be the *Archives of Sexual Behavior.*

INAUGURATION AND SURVIVAL STRATEGIES

To launch an effective sex research center, one must formulate a reasonably clear organization. There must be a governing body, a board of directors. This should follow the corporation model and consist of the senior research people plus individuals representing the host institution. This mixture provides a balance between detailed knowledge and personal involvement (the researchers) and a broader, more objective viewpoint (the host institution representatives). It also assures communication between the center and its host.

An advisory board without voting powers is often useful not only for advice but also for public relations. Thus a board may consist of persons with relevant specialized knowledge and also of persons influential in the host institution and community.

The distribution of offices and duties among the board of directors is determined by availability and aptitude, but reasonably clear descriptions of rights and obligations are a necessity to prevent ambiguity or overlapping responsibility. If possible, it is ideal to have an office manager-bookkeeper who is neither a board member nor a researcher.

There should be several grades of membership in the center, ranging from temporary part-time clerks or assistants on up to the board of directors. Job specifications and criteria for promotion should be clear and in writing. It is probably convenient to follow the policies of the host institution with a few amendments to meet the special needs of the center.

The center should formulate its own policies to which agreement is a requisite of membership. Vital policy topics include the following:

1. How money, space, clerical time, supplies, and equipment are to be allocated.
2. Who is empowered to make formal external affiliations or commit the center in terms of time or money.
3. Under what conditions a center member may speak for the center, and under what conditions he or she must speak purely as an individual.
4. What constitutes grounds for dismissal.
5. What channels exist for appeals from the decisions of supervisors or peers.
6. How the policy is delineated whereby the center rather than an individual has ultimate control over projects. The center should decide what projects it will sponsor, facilitate their completion, and — if necessary — replace principal investigators.

Since memories are fallible and easily distorted, it is wise to have an organizational manual which spells out structure, policies, and procedures in some detail.

A center should decide early in its career the extent to which it wishes to collect relevant materials. Will there be a library or some collection? Will the data generated by research projects be kept indefinitely and to whom are they accessible? Questions of this sort must be anticipated.

A center should also plan some public relations functions so as to be known and valued by the host institution and by the local community. While direct measurement of the benefits obtained from speaking to various groups and participating in panel discussions and conferences is difficult, such activities are unquestionably beneficial. At this point, I must add that consideration and patience with press and television pay dividends even though there inevitably will be some misquotation and distortion.

There is advantage in having the center a legal corporation or, if this poses problems with the host organization, having some small subsidiary corporate body outside the host institution. This permits the center to have some funds of its own to use freely. These unfettered center funds may be built up by the senior staff donating some portion of their lecture fees, royalties, or clinical charges, as well as by tax-deductible donations by persons outside the center.

Having broached the subject of money and support in general, I shall devote the remainder of this paper to the topic.

At a time when existing sex education and research organizations are struggling to survive, it may seem absurd to plan for new comprehensive centers and for an international consortium. Actually, survival may depend on these plans since they represent efficiency, a pooling of energies, a united front, and coordination. Sex research has much to offer society and science. Certainly Europe and North America are in the throes of rapid large-scale social changes in sexual attitudes and behavior which entail equally large-scale social, political, and emotional stresses and problems. Population control, female liberation, alternative marriage forms, individual freedoms vs. social controls — all such pressing issues directly relate to human sexuality. We can and should present a most persuasive argument to governmental agencies, private foundations, and universities to support the sex research which is necessary for society and the individual to satisfactorily deal with the problems inherent in sexuality, interpersonal relationships, and social control of sexual behavior. Indeed, one could even approach the business world for support, citing that sexual problems contribute to absenteeism both directly and through alcoholism and reduced efficiency. Certainly everyone in medicine piously states that a good sexual relationship is the cornerstone of mental and emotional health; it is time we asked that this dictum be backed by research support.

While we can make a very persuasive case for continued and expanded support, we must be able to present well-designed, economical plans which promise immediate and tangible results. The foundations and government are presently favoring problem-oriented rather than basic research.

In view of the nature of bureaucracy and the current financial situation, it is probably wiser to strive for a large number of part-time researchers in a center rather than for a smaller number of full-time persons. The part-time persons, thanks to their other duties, are more securely rooted in the host university, clinic, or hospital. Also, the concept of part-time research meshes nicely with the prevalent university ideal of all faculty being engaged in some research. Consequently it is possible to unobtrusively recruit part-time staff from other departments and gradually increase their commitment, ultimately trying for a half-time research appointment in one's center or, as second best, having the department simply reduce the person's teaching load to allow more research.

Another strategy aimed not at local host organizations but at foundations and government agencies is to have sex research centers submit joint grant applications. If effective coordination were assured, such applications should be far more attractive than individual proposals.

A certain amount of useful work can be obtained through teaching functions. I do not mean the traditional exploitation of students, but a *bona fide* learning process through participation in data gathering and data analysis. The amount of work may be moderate, as in the case of a term paper or research course, or it may be large, as in the case of a doctoral dissertation. If a sex research center were utilized by a number of departments as a source of material for dissertations, or if human sexuality were recognized as a field of concentration (either as a major or minor), the center could become an integral part of the university or medical school.

Sex research centers in hospitals and clinics can charge for their therapeutic services and those members who are making large salaries can make tax-deductible contributions to the center, contributions which might help hire an assistant or secretary for the donor.

The sex researchers in nonmedical situations are in a less favorable monetary position but still have untapped resources. Many organizations need and want various sorts of sex research information: for example, the American Medical Association, Veterans Administration hospitals, the American Psychiatric Association, the American Psychological Association, various marriage counseling groups, the American Law Institute, the National Library of Medicine, the Playboy Foundation, many insurance companies (including Blue Cross), and some pharmaceutical companies. There is not any reason why they should not pay for the information received.

Another source of revenue is the periodic presentation of programs or workshops which charge a registration fee. Such programs, known by a variety of titles, have become much more numerous in recent years. The better-known ones are those of the Institute for Sex Research, the Reproductive Biology Research Foundation, the Minnesota group, and the National Sex Forum, but there are easily a dozen others. These vary in length from a day to 10 days; some stress informational input while others emphasize attitude change. The range of subjects covered is broad. If properly run, these programs also serve to promote a good public image of sex research and advertise its value; if improperly managed, these programs can do our profession considerable damage since we are already suspected of using the name of science or therapy to arrange for our personal sexual gratification.

Foundations have always preferred the new and innovative to the old and routine, even though the old and routine may be performing a valuable service while the innovation is a gamble. Similarly, they find the concept of seed money more appealing than long-term support. However, I wish to suggest a way of making long-term support more palatable. A consortium of sex researchers could make a good argument for core-support grants to keep alive a general body of ongoing research from which the attractive new ideas and projects could grow. A number of foundations could agree to set aside some small part of their budgets for such core support and the total of this monetary pool would not only be substantial but would also assure the chosen comprehensive research centers of a dependable, even though modest, source of funds. The more foundations contributing to such a pool, the more acceptable sex research would become, and ultimately such timid giants as Ford might be induced to set aside some inconspicuous fraction of their budgets for this purpose.

If I seem to be obsessed with funding, the impression is correct. It is fashionable to talk about provocative ideas, new approaches, fresh viewpoints, and flashes of genius, but the fact is that research primarily advances through dogged and often dull, protracted work. There must necessarily be trial and error, blind alleys, and negative findings. All of this sums up to the truth that research requires much time and energy, that the researchers and their support staffs must be paid, and that without adequate funding little research is possible. Our first task is to organize, coordinate, and improve ourselves so as to not only obtain but also merit such funding.

Group Discussion

The problem of reconciling academic freedom with the need to conform to the policies of a center was discussed. Dr. Rubinstein suggested that this was not a matter of black and white, but of a balance between an individual's need for freedom and his need for the resources and protection which a center provides.

Dr. Green called attention to the role of the sex research center as a means of quality control for the work done under its aegis. Quality control is already a major issue in sex education and sex therapy; it should not be ignored in sex research.

Dr. Fordney-Settlage envisioned a university-linked but independent center which offered for one metropolitan area a package composed of sex education and sex therapy with a related sex research program — funded either publicly or perhaps by local industry.

Ethical Issues[1]

Following the presentation and discussion of individual papers, the last two Conference sessions were devoted to more general discussions. The first, devoted to ethical issues, is summarized below:

Dr. James led off the discussion by noting that her own professional concern with female criminals, drug addicts, and sexual privacy had sensitized her to a number of ethical issues. One is the issue of societal intervention in private lives — by means of laws, for example, determining what kinds of sexual acts are natural and licit, what kinds are unnatural and illicit. Laws such as these are an enormously controlling factor in human sexuality. What is their ethical basis?

Some ethical principles are quite simple — for example:

Any plan for research with human subjects should be reviewed in advance by a group independent of the group planning to conduct the research.

Human subjects in a research project should be protected from undue risk.

Personal data should be kept confidential. Scientists, she stressed, must depend on themselves to insure confidentiality, not on laws or institutions.

Other issues, Dr. James continued, are more complex — such as the ethical duty of a research scientist with respect to political activism. Her own feeling is that research scientists do not have an ethical duty to become social activists battling for the changes their findings support. Whether they will engage in activism in the arena of social change is a choice they are free to make. But, at the very least, even scientists who keep out of the arena themselves should support the rights of their colleagues who do elect to enter the arena and battle for social change.

Yet another ethical problem concerns the reporting of scientific findings which may be unwelcome to society or to a social group — or which may, indeed, be harmful to the status of a social group. Examples might be a finding that the disabled have little energy, that blacks have low IQs, that women are unstable, or that homosexuals are neurotic. What are the ethical implications of either publishing or suppressing such findings?

[1] Part of the proceedings of the conference, "Sex Research: Future Directions," held at the State University of New York at Stony Brook, Stony Brook, New York, June 5-9, 1974.

Dr. Rose made a statement concerning ethical issues, which is paraphrased below:

Scientific breakthroughs are rare; most scientific work is devoted to replicating prior findings and to expanding them. Hence there is a duty of scientists to express their hypotheses in ways which will facilitate their being refuted as well as confirmed. The sampling, the methodology, and the surrounding circumstances should be specified sufficiently to facilitate replication. Especially in the areas of education and therapy, results should be evaluated "blind" by someone other than the teacher or therapist.

The key ethical issues with respect to sex research on human subjects are informed consent and confidentiality. If normal adults are fully informed of the risks they are being asked to take, and if personal data about them are kept confidential, the ethical requirements relevant to sex research are in large measure satisfied.

National data registries and data banks raise a serious problem of confidentiality, since they can be (and have been) the source of abuse.

In New York State and no doubt elsewhere, there is a little-known law which states that the Commissioner of Public Health can grant confidential status, including immunity from subpoena, to research data. This law should be complied with and taken advantage of whenever confidential data are to be stored in a computer. In addition, sophisticated coding techniques — cryptography, if you will — should be exploited to prevent unauthorized access even if the computer tapes are stolen.

Dr. Rubinstein recalled that during one longitudinal study of college students where confidentiality was of great importance the coding key was stored in a Swiss bank, safe from either theft or subpoena. Dr. Rose added a further safeguard; the computer system should be set up so that data concerning one individual can be erased, if that becomes advisable, without simultaneously destroying the rest of the data.

One NIMH-funded project, it was reported, has the data which it gathers sent directly to Canada, where the names of respondents are recorded and the nameless data are then returned to the United States. This safeguards the names from subpoena by a United States court. Another approach described by Dr. Green is to deposit names in a safe-deposit box in another state, where they are immune from subpoena by the state where the data were gathered.

Dr. Cole expressed his distress at the need for such extraordinary safeguards, a need based on the feeling that sex is somehow more sensitive than other forms of human behavior. He expressed the hope that the day will come when sex taboos will melt, and the fear and horror of the subject will diminish, so that sex data can be handled like any other research data. Dr. Rose replied that precautions are needed not just for sexual data but for *all* personal data about a respondent. Drug data, for example, are as sensitive as sex data.

Dr. Gebhard pointed out that sex researchers, who associate mostly with one another, tend to underestimate the degree of bias and conservatism surviving in the real world. The great majority of individuals interviewed recently for the Kinsey Institute attitudinal study, for example, felt that a homosexual should not be allowed to teach school. Thus a teacher identified as a homosexual may lose his job. A marriage can similarly be destroyed by leaked data. Thus records must be safeguarded in every way possible.

Mr. Wiener called attention to the fact that, in addition to informed consent and confidentiality, other ethical issues were brought into focus by the Tuskegee research on syphilis, launched decades ago but only recently publicized. Here the ethics of withholding treatment from a placebo-treated group for decades was at issue. He predicted that if a Tuskegee-like proposal were submitted tomorrow, it would not be approved.

Dr. Davison stated that after some experience with aversion therapy he had become more and more unhappy with its use in the treatment, for example, of homosexuals, and had reached the point of being ethically opposed to any form of therapy designed to change homosexuals into heterosexuals. Some of his colleagues, in contrast, feel that changeover therapy should be an option available to homosexuals.

In buttressing a no-aversion-therapy position, he continued, it is easy to fall back on the supporting argument that aversion therapy doesn't work. This leaves one open to the dilemma, however, of what to do if a form of aversion therapy which *does* work should be announced. He has thus come to the conclusion that the argument of ineffectiveness is dangerous and inappropriate; he prefers to state bluntly that he is opposed to aversion therapy and other forms of changeover therapy for homosexuals, whether or not they work.

Dr. Green pointed out that in principle the psychiatrist or psychoanalyst does not set goals for the patient; the direction of the therapy is patient-activated. He doubted, however, whether the profession had fully lived up to that principle. Too often, the psychiatrist or psychologist approaches the patient or client with a preconception of what is good for him or her. Dr. Green expressed sympathy for the view that it is societal pressures which afflict homosexuals and which bring them in for therapy, and that there is need to treat the culture rather than the victims of the culture. Nevertheless, he suggested, in the concrete case where a homosexual comes to us asking for a therapy which will reorient him we may be dictating our own value judgments if we withhold therapy.

Mr. Wiener recalled a proposal some years ago in which aversive therapy would be made available to sex offenders who volunteered for it. The therapy involved a hundred painful electric shocks per minute. The sex offenders wanted this treatment; they wanted to change and to be released from prison. The review committee liked the goal of the project — producing change in sex offenders who wanted to be changed — but they were bothered by the aversive stimuli.

The review committee approved the project, but with so low a priority that it was never funded. Today, with the greater concern about human experimentation, such a project would in all probability go unapproved as well as unfunded.

A surgical parallel was suggested. Some surgery cannot be completed without pain. A patient who understands this may elect to suffer the pain to achieve the benefit, and may sign the surgical consent form. Refusing to inflict the pain is clearly wrong in the surgical case. Is the psychosexual case any different?

Dr. Bell pointed out one difference. It is easy to evaluate a consent for surgery. A desire to change sexual orientation, however, is much more complex. Some heterosexual clinicians hear only the request for change. It may turn out that the request for changeover is more immature and in need of therapy than the psychosexual status. Respect for the patient's motivations does not require instant response to whims or superficial verbalizations.

Dr. Goodman suggested that both approaches are flawed. A therapist should not be merely a technician who implements the patient's goals at the patient's request. Nor should the therapist dictate goals to another human being. The proper approach is for therapist and patient or client to reach informed agreement both on goals and on methods. This isn't a neat and simple solution, but solutions to ethical problems are rarely neat and simple. An ethical problem is rarely a matter of good vs. bad; typically it is a conflict between two positions, both of which have merits and defects. The therapist brings to the situation an expertise which the patient lacks; the patient brings information and goals. The two sets of inputs have to be brought together.

Dr. James suggested that the paternalistic approach of clinicians is a major factor in the problem. She also suggested that community activism, by altering the climate in which the patient exists, is an important adjunct of therapy. Community involvement of the research scientist is itself a contribution to therapy.

Dr. Green called attention to situations in which researchers are already involved in cases before the courts in which they are appearing as expert witnesses. He noted that he has appeared in cases involving child custody where mothers were lesbian, visitation rights of divorced fathers in cases where they were homosexual, issues of pornography and obscenity, and tests of the constitutionality of various common sexual behaviors (illegal) performed by consenting adults in private. The data we can bring to such cases are limited; but, even so, this is one direction in which sex research is already becoming an applied science within the arena of social change.

Dr. Bell described his own community involvement in Indiana, in forming a statewide coalition of gay liberation people, the National Organization of Women, Planned Parenthood, and the Indiana Council of Churches, meeting regularly in small groups, holding kaffeeklatches, publishing newsletters, in communities large and small. The goal is to increase people's consciousness of the fact that Indiana laws remain repressive. "We are going to be talking with legislators and others in an attempt to get people to know the facts and to open up dialogues between right-to-lifers and pro-abortionists, and so on."

Dr. James pointed out that research scientists can no longer speak in court from a philosophical base; what is needed are data. There is nothing more destructive than two scientists testifying in court on their conflicting versions of the truth on a pornography issue, for example, and having it all over the *New York Times.* A related problem is the criticism which a person in academic life gets from his or her own colleagues if he or she becomes involved in activism. Even a colleague who supports your research may criticize you for carrying your findings into the social arena. This situation is getting better; it will be remedied whenever academicians draw the distinction between getting directly involved themselves and supporting their colleagues who choose to become directly involved. If a research scientist becomes directly involved in changing smoking habits to prevent lung cancer or in urging ramps in public places for patients in wheelchairs, of course everyone applauds. The same attitude should be applied to research scientists who are trying to relieve the abuses perpetrated on certain sexual minorities.

Dr. Bell called attention to the ethical concern researchers should have for the effect of their reports on future researchers approaching the same population. Following a critical report on female homosexuals, for example, future researchers may find it hard to secure the cooperation of female homosexuals. But should findings critical of a population therefore be suppressed? By clarification of the issues, it was suggested, the results can often be presented in ways which are both honest and unlikely to give rise to hostility.

Dr. Cole felt that the research scientist should not feel responsible for how others use or misuse his findings. That is the ultimate in paternalism. One must retain a faith that human beings are capable of dealing with information. If they choose to misuse that information, it is their responsibility, not ours.

Mr. Wiener discussed a study of communes. The investigator found that the children in some communes were using drugs — marijuana and occasionally LSD but not heroin. The children were also engaging in sexual activities among themselves, and on occasion with adults. The investigator wondered if he should publish such data. If he did, there was danger that the communes in question would be damaged. After thinking about it and consulting lawyers, he decided to publish his results *without* names — and to go to jail if necessary for refusing to divulge the names.

Dr. Green called attention to the need for publishing data on such deviant forms of child-rearing. It might prove to be the case, for example, that children brought up in this way get all A's in school, enjoy educational and psychosocial developmental benefits, and so on. Society is entitled to such data and can benefit from it.

Dr. Green also described a problem which occasionally arises in his role as Editor of the *Archives of Sexual Behavior* when a manuscript goes against the grain of members of the Editorial Board. There are people on the board, for example, with very strong negative feelings toward behavior modification procedures with homosexuals. Even manuscripts which are methodologically sound have met with strong opposition on philosophical grounds — the view that be-havior modification for homosexuals doesn't belong in the scientific literature because of its political impact. This raises, Dr. Green suggested, First Amend-ment and censorship issues as well as the issue of the scientific ethic.

Another participant pointed out that research reports launched into a hos-tile social climate can often change that climate. The original Kinsey reports are striking examples. Society was wholly unprepared for them; the reports them-selves served to mitigate that hostility and open the door for further research.

Dr. Schmidt suggested that the conflict under discussion is a conflict inevi-table for scientists brought up in a value-free or value-neutral tradition as far as their scientific activities are concerned, but committed to values in the rest of their lives. He stated his own way of resolving this conflict. He accepts the fact that there is no value-free or value-neutral way to approach a problem in sex research. Nor should scientists attempt to approach this value-free ideal as closely as possible. The value-free ideal is useful only as a tactic for securing funding for a project. Sex research should in fact be directed toward the chang-ing of reality — the changing of public attitudes toward sex and the social impli-cations of sexual activities. This should be the main aim of sex research. He him-self, for example, is not investigating sexual attitudes out of idle curiosity, but to be better able to fight for appropriate goals. But in what direction should reality be changed? That is the truly important question, which each of us must answer for himself in approaching sex research.

Some researchers, Dr. Schmidt continued, begin with a definition of health, and then say they wish to change reality in the direction of better health. He regards this as too narrow and abstract. He prefers to say that he wishes to change reality in the direction of more social and sexual emancipation. This im-plies two things: as much sexual freedom as is possible in a specific social and historical situation, and liberating sexuality from being an instrument for exploi-tation of other people and for fostering individual or social dependency.

Dr. James pointed out that society strikes a balance between freedom and social controls. It is not for the research scientist to strike that balance, but rath-er to make available the relevant data and trust to its being rationally used. Sup-pressing any information goes against the whole scientific tradition.

Dr. LoPiccolo questioned this reliance on rationality. Any finding can be explained in many different ways; it is impossible to determine in many cases which explanation is rational and which ones are not. Simply providing data does not insure the rational interpretation of that data by others.

Dr. Lipman-Blumen noted that any research finding can be reported in many different ways, depending on the audience. Scientists write in one way for a professional audience and in another way for a lay audience. One of the audi-ences we must consider is the politically involved audience which may use the information for its own narrow political purposes.

Much depends, Dr. Lipman-Blumen added, on the way in which research is reported, or more commonly misreported, by the mass media. There is a need to educate reporters for the mass media to the point where they can report responsibly and accurately.

Dr. Green discussed the ethical issues raised in a recent California example involving rapists and child molesters held in an institution on indeterminate sentences, which can mean life imprisonment. It was proposed to release them after placing them on a drug, cyproterone acetate, widely used in Germany and more recently England, to reduce libido and sexual activity, with the full consent of the prisoners. Those who proposed this in California expected opposition from members of the public concerned that rapists and child molesters would be released and would constitute a public menace. Opposition came instead from civil libertarians who doubted the capacity of an imprisoned population to give informed consent without duress. The study was blocked, in part, by public opposition. In this case, how are civil liberties most threatened? Is it by "coercing" a prisoner to accept drug therapy — which will release him from prison — or by holding him prisoner indefinitely? Many other ethical issues in sex research involve similar choices among evils.

Concluding Discussion[1]

The last conference session is summarized below:

Dr. Messenger opened the final session of the conference by noting that repeatedly generalizations were made at prior sessions which had led him to think: "This is something which should be checked cross-culturally." The anthropological literature is filled with neglected data relevant to the topics here discussed. That neglect should not continue.

Dr. Gebhard agreed and cited two examples. Very few paraphilias are found in preliterate societies, as compared with our own. Why? In some societies there seems to be very little loss of libido with aging; what does this say about the causes of libido loss with aging in our society? It was also pointed out the reported incidence of homosexual behavior varies from zero in some cultures to 100% in others; what explains the divergence?

Dr. Messenger added that anthropologists through the decades have recorded sexual data which they have been unable or unwilling to publish. He recently proposed that this archival material be assembled and published, but nothing has come of the suggestion. One anthropologist has only recently published data which he collected 30 years ago.

Dr. Goy said that with respect to animal research the great need was for opportunities to undertake work which might cast light on those aspects of the human condition which are currently thought of as "problems." There are many parallels between the sexuality of the Old World primates and the sexuality of the human species. These parallels should be exploited for the light they throw on the human condition; and workers in human sexuality should consider the animal findings. Animal researchers can enter fields which are barred to human researchers by political or ethical considerations. Dr. Goy accordingly urged a marriage between the work of human researchers and that of primatologists. The need for that marriage, he concluded, was the point most clearly illustrated for him at this conference.

Dr. Lipman-Blumen called attention to the training of the next generation of sex researchers, and the adverse effect on that training of budget cutbacks and

[1] Part of the proceedings of the conference, "Sex Research: Future Directions," held at the State University of New York at Stony Brook, Stony Brook, New York, June 5-9, 1974.

changes in Federal training grant policies. Alternatives must be found for the predoctoral and postdoctoral training programs which have been cut off.

Dr. James raised the parallel problem of advanced degrees for graduate students interested in sex research. There are no such degrees in sex research *per se*, and it is very difficult in most institutions to put together an interdisciplinary Ph.D. program for graduate students whose chief interest is sex research.

Dr. Gebhard called attention to the conflict between the need for multidisciplinary research and the fact that few research agencies have representatives of many disciplines on their staffs. The consortium idea appeared to be one solution. Suppose, for example, that eight or nine scientists at as many institutions could be interested in participating in a particular sex research project. A multidisciplinary approach would result. It would also be useful if, after the data were collected, they were circulated to eight or nine specialists for comment and for reporting from various perspectives: "This is how the data look to me as a psychologist." "This is how they can be interpreted from a sociological point of view." Dr. Lipman-Blumen added that the same pattern might apply to researchers from different cultures, reporting data collected from each culture.

Interdisciplinary research, Dr. Rose noted, is for various reasons not always practical. But some of the benefits of interdisciplinary research can in almost any situation be readily achieved simply by having associates in other disciplines review a research protocol before the project is finalized. Short of true collaboration, this advance input from those of another discipline can contribute greatly to improving research design.

For obvious reasons, much human sex research is concerned with interview data, psychometric self-evaluations, and other verbal materials. Including biological, physiological, and behavioral data in a study can go a long way toward validating the verbal responses — even though the most direct and sophisticated measures of physiological and behavioral response may not be feasible in human research. Even simple measures will help. In retrospective studies, for much the same reason, self-recall reports should be supplemented by school records or other contemporary evidence as a means of validation.

Where human subjects are questioned, Dr. Rose continued, it is important to control for *their* concept of what you are expecting to prove. If this cannot be controlled for, it should at least be evaluated. The need for informed consent from subjects is related to this need to avoid contaminating the findings by the "set" instilled in subjects when their consent is solicited.

Dr. Geer urged that sexual behavior be viewed as a segment of all behavior, not independent of other parameters. Sexual behavior should be recognized as a socially conditioned act closely related to other psychological, social, and biological issues. Human sex is primarily in the head, primarily cognitive in nature. If you instruct a subject in the laboratory to "Think of a sexual fantasy and turn yourself on," something obviously happens. Changes occur in the autonomic nervous system, and changes occur in what the person is experiencing, without the presentation of any external stimuli. This "sex in the head" is ordinarily labeled fantasy; but we shouldn't stop there. Fantasy includes memory, recall of past events, recall of past fantasies, visual images, verbalizations, and so on — all hidden under the term "fantasy." To get below the term we don't need more sex research; we need communication with colleagues who have been studying fantasy. The same applies to other sex research problems which cannot be solved within the sex research field. Thus sex research must go beyond mere sex and must correlate itself with the rest of the human field.

Dr. Schmidt urged that in considering multidisciplinary approaches the approach of the social historians should not be overlooked. It is they who have made the major contributions, from a theoretical point of view, to sex research in Europe over the past 5 or 10 years. Reviewing European social history from medieval times to the present, they have made important findings concerning sexual behavior which should profoundly influence sociological and psychological research. Recent studies of social history have shown, for example, that religion has played a relatively minor role in European sexual repression; industrialization, by shaping a different personality structure, has been much more responsible for sexual repression. Every sex research center would be fortunate to have a social historian contributing to its work.

Dr. LoPiccolo expressed a concern that with questionnaires we are studying checkmark behavior rather than sexual behavior. Often, too, the presence of the observer or researcher alters the experiment. The ideal situation, accordingly, would be one in which unobtrusive measures are used, measures of a type which do not alert subjects to the fact that an experiment is going on. But this approach is blocked by the informed consent requirement.

Some sexual behavior issues, Dr. LoPiccolo continued, may be inherently unresearchable in the present climate of opinion; issues involving the status of women are examples. A study of sexual frustration in women, for example, found that women activists in the women's movement are less frustrated than other women in the sample. All this may mean, however, is that women in the women's movement are dedicated to persuading inquirers that they are not sexually frustrated.

In designing that study, Dr. LoPiccolo also noted, he and his associates did not first recruit women's activists and then match them with a control group. Such matching is always suspect, for the matching criteria may be irrelevant to the behavioral determinants. Instead, they just recruited women and determined their activist status at the same time as their frustration status was determined — a preferable strategy for assuring that the control group is comparable to the subject group.

In sum, Dr. LoPiccolo concluded, sound methodology in the social sciences requires methods — studying subjects in their natural environment, keeping both the researchers and the research measures so unobtrusive that subjects are unaware they are being studied. This approach will insure that we are studying reality rather than creating it. But it is precisely such methods which raise serious ethical problems.

Others pointed out that animal research does not require informed consent. Anthropologists in the past did not seek informed consent, but that is changing.

Dr. James called attention to the need for what she called "multiple convergent measures" in supporting a finding. In her study of prostitution, for example, she combined statistics, linguistic analysis, ethnology, and psychology. Had she relied on any one of these approaches, her results would have been both misleading and trivial.

Dr. Davison pointed to the need to report negative findings — findings which fail to confirm a hypothesis, for example, or which fail to find a difference between two groups when a difference is anticipated. Thus in some Ph.D. programs failure to confirm a hypothesis may lead to rejection of a Ph.D. thesis. In fact, failure to confirm is also a worthy finding — a point that should be borne in mind both by Ph.D. committees and by the editors of scientific journals.

Dr. Green briefly discussed the role of his *Archives of Sexual Behavior.* It is firmly based on an international, interdisciplinary board of editors who represent the major disciplines with significant input into sex research — animal, psychological, sociological, psychiatric, and so on. It is open to negative as well as positive findings. In editing the *Archives,* he has become acutely aware of the need for international collaboration and especially for international awareness of sex research findings. He has found excellent papers in Russian, Italian, and so on for translation and publication here — material which simply does not enter into the awareness of American scientists. We must learn to collaborate better across national boundaries and across languages. There may be salient research under way in the Orient, in Latin America, and elsewhere of which we are not aware. A modest flow of manuscripts from some of these areas to the *Archives* has begun.

Dr. Yolles called attention to a highly successful interdisciplinary research program involving groups at as many as 20 geographically separated centers. The object was to evaluate psychopharmacological agents. Not only were reliable evaluations generated, but also the cooperating groups then became work groups for defining policy with respect to these drugs. Out of the project grew an ongoing information program involving bulletins, a clearinghouse, and so on. He stated that he was urging, both within NIMH and elsewhere, the adoption of a similar pattern in the field of human sexuality, including research, training, and service components. Each of these components is greatly strengthened if the other two are included in a program with it. Funding is needed for all three in sex research centers throughout the country; but it also needs a focal point in Washington. Both applied and basic research should be included, and commitments should be for more than 2 years. The 2-year research commitment has damaged research in the United States to a remarkable degree. Five-year support is good; repeated 5-year grants are needed for basic groups doing basic and continuing research.

Dr. Yolles suggested that the prime audience which scientists should address for increased funding is the legislators and Congress. There has been a tradition against scientist involvement in the legislative arena, a feeling that it is demeaning — especially for the medical profession. The physicians have now learned the hard way that they *have* to address the audience of politicians. Academicians still shy away from appearing at Congressional hearings, however. A caveat here is to talk about what you know, and to have data. Lawmakers really are interested in getting informed opinions, but they need it in terminology they can understand.

Mr. Wiener said that he would have liked more data at this conference on several topics. Childhood gender roles had been explored, but childhood sexuality was rarely mentioned. The implications of animal research for human studies had been discussed; he would have liked also a discussion of the implications of human work for animal research.

Dr. Green observed that, despite its limitations of time and topics, this conference had been the most broadly based, stimulating, and erudite meeting on human sexuality he had attended. He thanked all participants on behalf of all four codirectors of the workshop, Dr. Yolles, Dr. Gebhard, Dr. Rubinstein, and himself.

Summary and Recommendations[1]

The Stony Brook Conference on the Future of Sex Research was concerned with three major topics:

1. The subject matter for future sex research.
2. The methodology of future sex research.
3. The ethics of sex research.

We can summarize what emerged from the conference on the second and third of these topics in very brief terms:

The methodology of sound sex research has few, if any, unique characteristics. The same scientific methods which have proved their worth in the social, psychological, and biological sciences generally are also the scientific methods effective in sex research. The same ethical issues arise in this field as in the other behavioral sciences where human beings are the usual subjects for research, and are soluble in much the same ways.

METHODOLOGY

The papers presented were based on a very wide range of methodological approaches. Yet none of these approaches, so far as we can see, differs in any essential respect from approaches familiar in the other sciences.

Goy's paper, for example, was concerned with the etiology of bisexual behavior and reported that such behavior can be elicited in adult rhesus monkeys if their exposure to sex hormones has been altered by experimental intervention during fetal development. We cannot run such an experiment on humans, of course — but this is hardly a limitation unique to sex research. Scientists in many fields are accustomed to run on nonhuman species a vast array of experiments which cannot be run on the human species.

Goy went on to describe both similarities and differences of response to experimental intervention among guinea pigs, rats, and rhesus monkeys — but this, too, is a common approach in comparative zoology and other sciences.

[1] Part of the proceedings of the conference, "Sex Research: Future Directions," held at the State University of New York at Stony Brook, Stony Brook, New York, June 5-9, 1974.

Goy called attention to ways in which animal models — specifically non-human primate models — can be used for studies of human homosexuality, bachelorhood, masturbation, impotence, and no doubt a wide range of other sexual phenomena common to both species.

We agree, and we urge generous funding of projects based on experimental intervention in nonhuman models.

Any study of a nonhuman species, of course, gives rise to the ubiquitous question: Is this also true of humans, or, more precisely, to what extent is this also true of humans? Here the work of Money, Green, and others calls attention to another potentially very fruitful methodological approach: the study of "experiments of nature," We cannot directly intervene to alter the endocrinological milieu of the human fetus, but in certain well-defined syndromes nature itself alters the fetal milieu — or it may be medically altered, as when hormones prescribed therapeutically for the mother pass through the placenta. Longitudinal studies of humans thus exposed to atypical hormonal milieu during fetal development are helping to determine whether what is true of rhesus monkeys following experimental intervention is also true of the human where parallel intervention is prohibited.

The anthropologists present at the conference — Gebhard, James, and Messenger — called repeated attention to the potential usefulness of cross-cultural studies. They can be useful in at least two ways. Where a form of human sexual behavior appears to vary widely from culture to culture (e.g., homosexual behavior, which is reported to vary from zero in some cultures to 100% in others), we can search for the cultural correlates of the reported frequencies. Conversely, where a form of sexual behavior is quite uniform across cultures, we can avoid the trap of attributing this form of behavior to factors unique to one culture. In none of these respects, however, is sex research different from other behavioral sciences. Cross-cultural studies, like cross-species studies, are a commonplace of the behavioral sciences generally. The remaining resistances and taboos impeding research on human sexuality in our culture make cross-species and cross-cultural approaches particularly inviting.

Perhaps the dominant theme of the conference, as far as methodology is concerned, was the repeated recognition of the need to approach questions using a variety of converging methodologies rather than seeking that will-o'-the-wisp, the ideal methodology. The conjoint use of retrospective and prospective approaches, of longitudinal and cross-sectional approaches, is a striking case in point. Green for example, reported on his longitudinal, prospective study in which children who exhibit an atypical sex role identity and atypical sex role behavior are studied at an early age and are then followed to determine adult outcomes. Bell, in contrast, reported on a large-scale retrospective, cross-sectional survey of males and females engaging in homosexual activities in one large American city. Both studies have obvious limitations. Since Bell must rely on subjective recollections of what happened to his subjects years or decades ago, his study is subject to the vagaries of subjectivity and impaired recall. Green's study must rely on the few subjects in his control group to tell us about those cases of atypical sexual behavior which are *not* identifiable at an early age. By combining these quite disparate approaches to atypical sexual behavior, however, we can derive a much fuller picture of that behavior than either study by itself can provide. To the extent that the findings of the two studies confirm one another, our confidence in both will be enhanced. Conflicting findings may pinpoint problems which will in turn become fruitful opportunities for further research.

Much the same is true of objective and subjective methodological approaches to a problem. Thus Schmidt asked respondents for a subjective report on their feelings of sexual arousal during exposure to various types of stimuli. Geer and others pointed out that arousal during such exposure can also be studied by objective physiological techniques such as penile and vaginal plethysmography. Schmidt's approach provides data on the subjective parameters of sexual arousal, while a plethysmographic study might obtain data unavailable by Schmidt's technique (such as the precise timing of response onset and magnitude). To the extent that a plethysmographic study confirms Schmidt's findings, it would confirm the reliability of Schmidt's respondents; discrepancies between the two studies might lead to further research seeking to trace them to inaccurate subjective reporting, to technical artifacts (such as the alteration of responses by the presence of the plethysmograph), or to differences between the subjective and objective parameters of sexual arousal.

The *replication* of findings is, of course, a methodological commonplace in many scientific fields; yet doing the same work over again is costly and usually yields information of only limited value. No one wants to be the *second* scientist to discover a phenomenon. Ways around this need for and dislike of replicational studies were discussed in the course of this conference. Bell, for example, embedded in his San Francisco questionnaire instrument some questions which had also been asked in earlier studies in other cities and in studies of national samples. Thus his study constitutes a partial replication of earlier studies, yet also provides fresh data.

On this point, Schmidt offered a methodological caveat. Vast areas of human sexual behavior and response, including areas of immediate practical importance, remain wholly or almost wholly unexplored. Manpower, funds, and facilities devoted to teasing out minor weaknesses in prior studies of well-explored areas reduce the pool of manpower, funds, and facilities available for exploring new realms. The allocation of resources to fresh fields of inquiry, to follow-up expansions of prior studies, and to the replication of prior studies is a matter requiring sound judgment on the part both of those who draft research proposals and of those who pass upon them. In the present posture of sex research, a field which is barely beginning to come into its own, we confess a modest prejudice in favor of innovative approaches, with replication projects limited to cases where there is a clear need. Often, when the findings of two studies are in conflict and replication therefore appears necessary, a fresh approach may be found which will both resolve the conflict and provide valid data or insights. Unlike such well-developed fields as astronomy or physics, sex research is not yet in a position to afford the luxury of replication for replication's sake.

There are obviously exceptions to this generalization, as is superbly illustrated by Fordney-Settlage's paper at this conference. Superficially, she is engaged in replicating the program of sex therapy initiated by Masters and Johnson in St. Louis. The population to whom she is offering sex therapy, however, is startlingly different from the well-educated, largely white, largely middle-class Masters and Johnson sample. As a result, she is able to report on problems quite different from those brought by patients to Masters and Johnson, and has had to develop approaches which differ notably in some respects from those developed in St. Louis. That some Masters and Johnson findings are confirmed in so disparate a population constitutes partial replication — but signifies far more.

We must learn to crawl before we walk or run. It is easy to forget this truism in an underdeveloped field like sex research, which operates in an academic setting surrounded by more prosperous scientific fields. Lewis's paper at the conference is a reminder. Lewis calls attention to the need simply to observe what is actually going on — when a mother picks up her baby for the first time, for example, or when a father comes home from work and glances (or fails to glance) at his baby in the cradle. The anthropologist who first stumbles on a preliterate society which has had no prior contact with civilization is not expected to launch complex and well-controlled ethnographic or statistical studies; the first duty is to look around and report what he or she observes. There is similarly a need for scientists exploring fresh aspects of human sexual behavior simply to observe and report before they study or intervene or develop and confirm hypotheses.

Funding, of course, is intimately related to methodology. If adequate funds and manpower were available, much might be learned by drawing a random sample of a few thousand infants and following them longitudinally through the decades, using all of the methods known to sex research. Lacking adequate funding, we must use the best available strategies for studying small selected samples longitudinally, groups of moderate size intensively, and random nationwide samples to the extent that funds permit. By coordinating disparate approaches, we may develop a coherent picture of human sexual behavior, its antecedents and consequences. In the absence of coordination, we shall be left with isolated bits and pieces.

At several points, the proceedings of this conference reveal the importance of taking advantage of unplanned and unanticipated research opportunities. The fact that certain diabetic women years ago were given sex hormones during pregnancy, for example, provided an opportunity to study the effects of prenatal hormone exposure in humans — an opportunity which might easily have been overlooked in the absence of an alert scientist. Similarly, a study outside the field of human sexuality may, if modified only slightly, yield at modest cost data of value to sex research; the addition of sexual questions to a nationwide interview study focused primarily on other matters is a case in point. We urge a continuing alertness and funding of such strategies, especially because they offer a very favorable cost-effectiveness ratio.

We again stress that the scientific methods appropriate to sex research differ only in detail from the similar methods in common use in the other behavioral sciences. Good science is good science, whether the subject is human sexuality or human gastrointestinal functioning. Perhaps an important difference between this and other fields is the even greater reliance we must place on following a variety of disparate methodological approaches, and on coordinating those disparate approaches so that they will gradually produce a mature science.

ETHICS OF SEX RESEARCH

As in the case of methodology, there is little in the ethical field which differentiates sex research from the other sciences generally, and especially from the other behavioral sciences.

Participants in this conference stressed three major ethical issues:

1. The need for confidentiality.
2. The need for informed consent.
3. The need for protecting research subjects from harm.

These are also, of course, the major ethical issues confronting research in the other behavioral sciences. Wtih respect to confidentiality, for example, even the routine information collected during the decennial census is by law held inviolate. Research subjects have a right to have *all* personal information kept confidential — financial and drug-use information and even age and national origin, as well as sexual information. It remains true, however, that sexual information is in some respects peculiarly sensitive. Indeed, the study of sexual behavior, because it is emotionally laden and subject to social pressures, may produce particular problems relating to the protection of human subjects. The knowledge that a respondent has engaged in homosexual or extramarital behavior, for example, may cause loss of a job or destroy a marriage.

Conference participants pointed out in this respect that the establishment of computerized data banks constitutes a threat to research confidentiality, and discussed ways of minimizing that threat. They called attention to laws in some states guaranteeing the confidentiality of research data — but stressed that research scientists must themselves bear primary responsibility for safeguarding their own data rather than relying on laws, review committees, or other outside aids.

With respect to all three ethical principles — confidentiality, informed consent, and protection of research subjects from harm — it was generally agreed that research protocols should be reviewed in advance (as they now are reviewed) by committees of the sponsoring and funding institutions.

It was also noted, however, that the potential harm which may be done to research subjects must be balanced against the potential good. Research aimed at facilitating the safe release of prisoners from prolonged incarceration is a dramatic case in point. The potential harm done to research subjects must be balanced against the harm done by prolonged, perhaps lifelong, imprisonment in the absence of the research. Indeed, the principle of informed consent could be meaningless if all potentially harmful research is to be banned.

Ethical problems involved in the *publication* of research findings were also raised at the conference — such as the question of suppressing findings which may be harmful to society or to a minority group within society, and the question of a scientist's responsibility to battle for the social changes which his research findings indicate are needed. These are problems met within other behavioral research fields as well. The maintenance of high ethical standards here is of special importance. There remains a lingering distrust of sex in our culture, and this distrust extends to sex research. A single violation of ethical principles anywhere tends to be generalized into an attack on all sex research. Constant vigilance against a violation of ethical principles is a price we should willingly pay for continued public confidence and support.

SUBJECT MATTER FOR FUTURE SEX RESEARCH

It was not our intention, when designing this conference, to emerge with a "shopping list" of sex research projects which should be promptly funded and launched — nor is that the intention of this summary. Rather, we here present a list of seven guidelines to aid in the weighing of future sex research proposals.

The first five of these seven guidelines are all directly concerned with problems in the world outside our laboratories — problems which are currently perceived as pressing. From the almost unlimited range of possible topics for future sex research, we recommend that primary emphasis be placed on these "problem-oriented" areas, with equal priority given to all five. In identifying these five areas we recognize that a number of others might have been included. However, the areas described were those discussed in greatest detail at the conference.

Area 1: Sexual Function Among the
Physically Handicapped

The first area is that in which Cole and Geiger work; they are concerned with sexuality among the physically handicapped. Cole estimates that as many as 10% of the total population of this country have a physical handicap which poses a substantial limitation on normal activities — including in many or most cases sexual activities. Cole's examples include arthritis, stroke, heart disease, end-stage renal disease, amputations, deformities, visual and auditory impairment, disfiguring scars, paralysis, and developmental anomalies. If the effects of aging are added in, almost all of us are at risk of physically based curtailment of sexual function sooner or later. The crippling and disheartening effects on the quality of human life, and the possibility of maintaining or restoring sexual function even in the presence of such extreme handicaps as paraplegia, were documented.

Area 2: Studies of the Sexual Minorities

A similar high priority should be placed on research projects concerned with the sexual minorities, among whom homosexuals constitue the largest single group. Also included, however, are the whole range of other forms of sexual variants.

One reason for placing these two groups — the sexually disadvantaged and the sexually atypical — at the head of the list, of course, is simple concern with human suffering. In addition, however, we are convinced that research concerning the sexually disadvantaged and the sexually atypical can illuminate the basic nature of sexuality in the human species. Helmholtz once remarked that the entire functioning of the human eye could be inferred from a study of the various forms of color-blindness. The study of how and why some children grow up to be homosexuals or fetishists or celibates can similarly teach us much about how and why most children grow up to be heterosexuals.

Area 3: Studies of the Therapy of Human
Sexual Dysfunction

Deserving similar priority are projects related to the treatment and cure of sexual dysfunction. Masters and Johnson in 1971 announced a fresh approach to the group of problems included under this rubric, and the treatment of sexual dysfunction has since then burgeoned in many centers; but demand still vastly exceeds supply. Publicity in the mass media concerning sex therapy has produced a revolution of rising expectations. Many questions remain unanswered — such as the evaluation of short-term and long-term benefits (including nonsexual benefits) for various types of dysfunction therapy in various populations, more economical methods of delivering services so that their availability will not be limited to affluent patients, improved therapeutic techniques, and more.

Area 4: Gender Role and Gender Identity Studies

One of the women participants in this conference, Lipman-Blumen, repeatedly called attention to parallels between the position of women in our culture and the position of the physically handicapped and of the discriminated-against sexual minorities. Indeed, at one point, Lipman-Blumen ventured to include being female among the physical handicaps in our culture. The rise of the women's movement during the past few years has served to underline both the handicap under which women in our culture function and the unwillingness of a growing minority of women to continue under that handicap.

For sex research, these developments have led to a whole series of considerations which occupied a major segment of the time of the Stony Brook conference. Just how do females differ from males? To what extent are they biologically determined or influenced? How can gender differences and gender role differences be altered, and what will be the effects of altering them? Five of the nine papers at the conference — those of Lewis, Green, Schmidt, Goy, and Lipman-Blumen — were devoted at least in part to explorations in this broad and emotion-laden area.

We see this emphasis as fully warranted, and urge that research projects concerned with women and women's role and identity in our society be given equally high priority with projects concerned with the sexuality of the physically and mentally handicapped, the sexual minorities, and the sexually dysfunctional. We recommend parallel studies of male gender roles, gender identity, and sexual function both for their own sake and for the light they can cast on female-male differences.

Area 5: Studies of the Changes in Contemporary Mores

Sexual mores are changing in our culture — including laws, attitudes, and practices concerning pornography, premarital sexual relations, alternatives to monogamy, contraception, abortion, and special variants. The fact that these issues are oridinarily debated in moral rather than scientific terms does *not* remove them from the subject matter of science. Understanding a phenomenon is a basic prerequisite for making informed moral judgements. Thus, while science cannot answer moral questions, it can and should supply the knowledge and understanding on which sound moral judgments can be based. Not only lawmakers but also the public at large need factual data in these morally sensitive areas, and we see the supplying of those data as a priority duty of sex research.

The remaining two areas to which we assign a high priority involve basic rather than problem-oriented research. No science, it is quite generally agreed, can flourish if it concentrates wholly on topics currently conceived as being "problems," to the neglect of its own theoretical underpinnings.

Giving priority to the five problem areas identified above, we believe, will also incidentally produce new knowledge in basic theoretical issues. The fact that some paraplegics report the sensation of orgasm despite a severed spinal cord, for example, arises out of problem-oriented research, yet affords important information concerning the nature of the orgasmic experience. Green's longitudinal study of the development of very feminine boys similarly throws light on theoretical aspects of human development in general. To the extent that the theoretical yield of a research proposal can be gauged in advance, accordingly we recommend that preference be given to those problem-oriented projects which give promise of also yielding theoretical insights. We are not content, however, with reliance on such contributions to basic research into human sexuality. We believe that the following two areas of basic research warrant equal priority with the five problem-oriented areas:

Area 6: The Neurophysiology and Biochemistry of Sexual Response

One wit has remarked that "so far as sexual physiology is concerned, Masters and Johnson have dragged us, kicking and screaming, headlong into the nineteenth century." One need only compare the vast methodological resources of contemporary neurophysiology and biochemistry with the crude observational data on which we still rely for our understanding of the human sexual response cycle to appreciate the justice of that remark. To cite a simple but crucial example, almost nothing is known of the neurological and biochemical antecedents of orgasm in either the human female or the human male. What little we do know is expressed in such gross and inadequate terms as muscle tensing and blood vessel engorgement, typical nineteenth-century concepts. So far as infrahuman species are concerned, we do not even know whether orgasm or orgasmlike responses occur in the female. Until we know at least as much about sexual functioning as we know about gallbladder functioning or auditory functioning, we must stagger about in the dark when designing problem-oriented research projects. We accordingly recommend the same high priority for these basic studies as for the five problem-oriented areas described above — not just to satisfy idle curiosity, but to establish a sound theoretical foundation for future problem-oriented research.

Area 7: The Unforeseeable

If the six guidelines for future sex research which we have set forth above were adhered to, the Kinsey research or the Masters and Johnson response-cycle research might never have been funded. On reflection, however, we are of the opinion that this is inherent in the nature of guidelines and of breakthrough scientific advances. Guidelines are concerned with the foreseeable; the breakthrough study penetrates a new region whose richness was not and could hardly have been foreseen.

Our final category of priority studies, accordingly, is reserved for studies which fit none of the six categories above but which are nevertheless destined to revolutionize our understanding of human sexuality. More realistically, we reserve this category for gambles — for research projects which depart from the mainstream of scientific interest and violate existing rules of project selection.

Funding the "odd-duck" idea involves a risk that funds will be wasted. Refusing to fund it involves the risk that basic progress will be delayed — and that the funding agency will have missed a unique opportunity. We recommend that when considering innovative proposals, however unorthodox they may appear, sponsoring and funding agencies consider both risks as we search for greater sexual knowledge and health.

Epilogue[1]

Every once in a while a scientific workshop combines substantive interchange with a high order of joyful cameraderie. When this occurs with a group working together for the first time, the effect is both productive and pleasurable. The following summary of the highlights was quite literally stimulated not just by the flow of ideas over the first 3 days of meeting but by the manner in which all participants quickly became active partners in the intellectual interchange. The humorous tone of the poem is the only way the author could convey this happy exchange. The poem was read to the group at the beginning of the last morning to provide a stimulus for the summary discussion that followed. While the content has a special meaning to those to whom it was addressed at the meeting, it is hoped that the reader will understand that the rhyming is incidental to the reasoning.

NIMH, I'VE GOT A LITTLE LIST

As some day it may happen that a project might get paid.
    I've got a little list — I've got a little list,
Of researchable ideas we all have just surveyed,
    And that cannot be dismissed — they must not be dismissed!

Are the limits of plasticity ultimate or not?
And is bisexuality biological, or what?
Can genes protect against the androgen effect?
How valid are the facts when data retrospect?
Is the incest barrier biologically based;
Or just a socially determined matter of distaste?
    And there's the socioemotional developmentalist.
    He cannot be dismissed, she must not be dismissed!

[1] Part of the proceedings of the conference, "Sex Research: Future Directions," held at the State University of New York at Stony Brook, Stony Brook, New York, June 5-9, 1974.

Is the head where the hard is, or what is reality?
And how about that old heredity-environment duality?
Is the vagina nature's own penile plethysmograph?
Or is that just a jest on a sexologist's epitaph?
If the man is too fast, is the woman too slow?
Or is role reversal the better way to go?
    What we need is a philosophical gynecologist.
    She cannot be dismissed, he must not be dismissed!

There's the child who recognizes what the adult cannot place;
The sex of other children from pictures of the face!
There's the question of the comprehensive center,
Can it help to make sex research excellenter?
Is homosexuality singular or plural?
And is it more abundant in the urban or the rural?
    And what about the multimetric methodologist?
    He cannot be dismissed, she must not be dismissed!

Is it better to be put down while you're also turning on?
Or would you rather equalize and find the feeling gone?
Are males more predisposed to others of same sex?
Or is that homosocial view a mirage quite multiplex?
Is impotence the plight of the finickiest male?
Or is that rhesus thesis just another fairy tale?
    We need an ethological endocrinologist.
    She cannot be dismissed, he must not be dismissed!

What happens to a child whose father was his mother,
Before dad changed one set of genitals for another?
There were many other issues cleverly refined,
Such as — What d'ye call it — Thing'em bob, and likewise — Never mind.
And 'tsk-'tsk-'tsk and That approach and also You know who —
The task of filling up the blanks I'd rather leave to *you*.
    But it really doesn't matter what you put upon the list,
    For they cannot be dismissed — they must not be dismissed!

Eli A. Rubinstein
June 9, 1974

References[1]

Acker, J. (1973). Women and social stratification: A case of intellectual sexism. In Huber, J. (ed.), *Changing Women in a Changing Society*, University of Chicago Press, Chicago, pp. 174-184.

Aldous, J. (1969). Occupational characteristics and males' role performance in the family. *J. Marriage Family* 31(4): 707-712.

Andreas, C. (1971). *Sex and Caste in America*, Prentice-Hall, Englewood Cliffs, N.J.

Baker, H., and Stoller, R. (1968). Can a biological force contribute to gender identity? *Am. J. Psychiat.* 124: 1653-1658.

Bart, P. (in progress). *A Study of Divorced Fathers*, University of Illinois Medical School, Chicago.

Beach, F. A. (1942). Execution of the complete masculine copulatory pattern by sexually receptive female rats. *J. Genet. Psychol.* 60: 137-142.

Beach, F. A. (1968). Factors involved in the control of mounting behavior by female mammals. In Diamond, M. (ed.), *Perspectives in Reproduction and Sexual Behavior*, Indiana University Press, Bloomington, pp. 83-131.

Beach, F. A. (1970). Hormonal effects on socio-sexual behavior in dogs. In Gibran, M., and Plotz, E. J. (eds.), *Mammalian Reproduction*, Springer, Berlin, pp. 437-466.

Beach, F. A. (1971). Hormonal factors controlling the differentiation, development, and display of copulatory behavior in the ramstergig and related species. In Tobach, E., Aronson, L., and Shaw, E. (eds.), *The Biopsychology of Development*, Academic Press, New York, pp. 249-295.

Beach, F. A., Kuehn, R. W., Sprague, R. H., and Anisko, J. J. (1972). Coital behavior in dogs. XI. Effects of androgenic stimulation during development on masculine mating responses in females. *Horm. Behav.* 3: 143-168.

Bell, A. P. (1974). Homosexualities: Their range and character. In Cole, J. K., and Dienstbier, R. (eds.), *Nebraska Symposium on Motivation: 1973*, University of Nebraska Press, Lincoln.

Bell, A. P., and Weinberg, M. S. (in progress). *The Development and Management of Homosexuality*, 2 vols.

Bell, R. R., and Chaskes, J. B. (1970). Premarital sexual experience among coeds: 1958 and 1968. *J. Marriage Family* 32: 81-84.

Benedick, T. G. (1971). Disease as aphrodisiac. *Bull. History Med.* 45: 322-340.

Benjamin, H. (1966). *The Transsexual Phenomenon*, Julian Press, New York.

Bensman, A., and Kottke, F. J. (1966). Induced emission of sperm utilizing electrical stimulation of the seminal vesicles and vas deferens. *Arch. Phys. Med. Rehab.* 47: 436-443.

Benson, L. (1972). *Fatherhood: A Sociological Perspective*, Random House, New York.

Biddle, B. J., and Thomas, E. J. (eds.) (1966). *Role Theory: Concepts and Research*, Wiley, New York.

---

[1] These references are cited in the papers presented at the conference, "Sex Research: Future Directions," held at the State University of New York at Stony Brook, Stony Brook, New York, June 5-9, 1974.

Bieber, I., *et al.* (1962). *Homosexuality: A Psychoanalytic Study,* Basic Books, New York.

Bors, E., and Comarr, A. E. (1960). Neurological disturbances of sexual function with special reference to 529 patients with spinal cord injury. *Urol. Surv.* 10: 191-222.

Brim, O. G., Jr. (1968). Adult socialization. In Clausen, J. A. (ed.), *Socialization and Society,* Little, Brown, Boston.

Brodie, H. K., *et al.* (1974). Plasma testosterone levels in heterosexual and homosexual men. *Am. J. Psychiat.* 131: 82-83.

Brooks, J., and Lewis, M. (1974). Attachment behavior in thirteen-month-old, opposite sex twins. *Child Develop.* 45: 243-247.

Brown, D. (1956). Sex role preference in young children. *Psychol. Monogr.* 70: 421.

Carnegie Commission on Higher Education (1973). *Opportunities for Women in Higher Education,* McGraw-Hill, New York.

Carpenter, C. R. (1942). Sexual behavior of free ranging rhesus monkeys (*Macaca mulatta* II): Periodicity of estrus, homosexual, autoerotic, and nonconformist behavior. *J. Comp. Psychol* 33: 143-162.

Carrier, J. M. (1971). Participants in urban Mexican male homosexual encounters. *Arch. Sex. Behav.* 1: 279-291.

Chafetz, J. S. (1974). *Masculine/Feminine or Human?* Peacock, Itasca, Ill.

Christensen, H. T. (1966). Scandinavian and American sex norms: Some comparisons with sociological implications. *J. Social Issues* 22(2): 60-75.

Christensen, H. T. (1971). *Sexualverhalten und Moral,* Rowohlt, Reinbek.

Christensen, H. T., and Gregg, C. F. (1970). Changing sex norms in America and Scandinavia. *J. Marriage Family* 32: 616-627.

Ciaccio, L. A., and Lisk, R. D. (1971). Estrogen: Effects on development and activation of neural systems mediating receptivity. In Ford, D. H. (ed.), *Influence of Hormones on the Nervous System: Proceedings of the International Society of Psychoneuroendocrinology,* Karger, Basel, pp. 441-450.

Clemens, L. G., and Coniglio, L. (1971). Influence of prenatal litter composition on mounting behavior of female rats. *Am. Zoologist* 11: 617 (abst.).

Cole, T. M., Chilgren, R., and Rosenberg, P. (1973). A new programme of sex education and counseling for spinal cord injured adults and health care professionals. *Int. J. Paraplegia* 11: 111-124.

Coleman, J. S. (chairman) (1973). *Youth: Transition to Adulthood, Report of the Panel on Youth of the President's Science Advisory Committee,* U.S. Government Printing Office, Washington, D.C.

Collins, R. (1971). A conflict theory of sexual stratification. *Social Probl.* 19: 3-20.

Comarr, A. E. (1970). Sexual function among patients with spinal cord injury. *Urol. Int.* 25: 134-168.

Cuber, J. F., and Harroff, P. B. (1966). *Sex and the Significant Americans,* Appleton-Century-Crofts, New York.

Current Population Reports, Population Characteristics (1972). Series P-20, No. 240.

Dahlstrom, E. (ed.) (1967). *The Changing Roles of Men and Women,* Beacon Press, Boston.

de Beauvoir, S. (1953). *The Second Sex,* Parshley, H. M. (transl.), Knopf, New York.

Diamond, M. (1965). A critical evaluation of the ontogeny of human sexual behavior. *Quart. Rev. Biol.* 40: 147-75.

Eaton, G. (1970). Effect of a single prepubertal injection of testosterone propionate on adult bisexual behavior of male hamsters castrated at birth. *Endocrinology* 87: 834-940.

Eaton, G., Goy, R. W., and Phoenix, C. H. (1973). Effects of testosterone treatment in adulthood on sexual behavior of female pseudohermaphrodite rhesus monkeys. *Nature New Biol.* 242: 119-120.

Edwards, D. A., and Burge, K. G. (1971a). Estrogenic arousal of aggressive behavior and masculine sexual behavior in male and female mice. *Horm. Behav.* 2: 239-245.

Edwards, D. A., and Burge, K. G. (1971b). Early androgen treatment and male and female sexual behavior in mice. *Horm. Behav.* 2: 49-58.

Ehrhardt, A. (1973). Maternalism in fetal hormonal and related syndromes. In Zubin, J., and Money, J. (eds.), *Contemporary Sexual Behavior,* Johns Hopkins University Press, Baltimore.

Ehrhardt, A., Epstein, R., and Money, J. (1968a). Fetal androgens and female gender identity in the early-treated adrenogenital syndrome. *Johns Hopkins Med. J.* 122: 160-167.

Ehrhardt, A., Evers, A. K., and Money, J. (1968b). Influence of androgen on some aspects of sexually dimorphic behavior in women with the late-treated adrenogenital syndrome. *Johns Hopkins Med. J.* 123: 115-122.

Ehrlich, G. E. (ed.) (1973). *Total Management of the Arthritic Patient,* Lippincott, Philadelphia, pp. 193-208.

Ehrmann, W. W. (1959). *Premarital Dating Behavior,* Holt, New York.

Eisenstadt, S. N. (1971). *Social Differentiation and Stratification,* Reiss, A. J., Jr., and Wilensky, H. (eds.), *Introduction to Modern Society Series,* Scott, Foresman, Glenview, Ill.

Epstein, C. F. (1971). *Woman's Place: Options and Limits in Professional Careers,* University of California Press, Berkeley.

Ernst, J. P., Gestefeld, M., Schulte-Westenberg, J., Seidensticker, M., Schmidt, G., and Schorsch, E. (1975). Reaktionen auf sexuell-aggressive Filme. In Schorsch, E., and Schmidt, G. (eds.), *Ergebnisse der Sexualforschung,* Kiepenheuer and Witsch, Cologne.

Finley, F. R. (1972). Research and training accomplishments in bioengineering. Paper presented at the Conference in Rehabilitation Research and Training Center, Philadelphia.

Firestone, S. (1970). *The Dialectic of Sex,* Morrow, New York.

Fox, C., Ismail, A., Love, D., Kirkham, K., and Loraine, J. (1972). Studies on the relationship between plasma testosterone levels and human sexual activity. *J. Endocrinol.* 52: 51-58.

Fox, J. (1971). Sex education — But for what? *Special Educ.* 60: 15-17.

Friedan, B. (1963). *The Feminine Mystique,* Norton, New York.

Gagnon, J. H., and Simon, W. (eds.) (1967). *Sexual Deviance,* Harper and Row, New York.

Gagnon, J. H., and Simon, W. (1973). *Sexual Conduct,* Aldine, Chicago.

Garai, J. E., and Scheinfeld, A. (1968). Sex Differences in mental and behavioral traits. *Genet. Psychol. Monogr.* 77: 169-299.

Gebhard, P. H. (1973). Sex differences in sexual response. *Arch. Sex. Behav.* 2: 201-203.

Goldberg, S., and Lewis, M. (1969). Play behavior in the year-old infant: Early sex differences. *Child Develop.* 40: 21-31.

Goldfoot, D. A., and van der Werff ten Bosch, J. J. (1975). Mounting behavior of female guinea pigs following prenatal and adult administration of the propionates of testosterone, dihydrotestosterone and androstanediol. *Horm. Behav.* (in press).

Goldfoot, D. A., Feder, H. H., and Goy, R. W. (1969). Development of bisexuality in the male rat treated neonatally with androstenedione. *J. Comp. Physiol. Psychol.* 67: 41-45.

Goslin, D. A. (1969). *Handbook of Socialization Theory and Research,* Rand McNally, Chicago.

Goy, R. W. (1968). Organizing effects of androgen on the behaviour of rhesus monkeys. In Michael, R. P. (ed.), *Endocrinology and Human Behaviour,* Oxford University Press, London, pp. 12-31.

Goy, R. W., and Phoenix, C. H. (1971). The effects of testosterone propionate administered before birth on the development of behavior in genetic female rhesus monkeys. In Sawyer, C., and Gorski, R. (eds.), *Steroid Hormones and Brain Function* (UCLA Forum in Medical Sciences, No. 15, Chap. 19), University of California Press, Berkeley, pp. 193-201.

Goy, R. W., and Resko, J. A. (1972). Gonadal hormones and behavior of normal and pseudohermaphroditic nonhuman female primates. In *Recent Progress in Hormone Research, Vol. 28: The Proceedings of the 1971 Laurentian Hormone Conference,* Academic Press, New York, pp. 707-733.

Goy, R. W., and Young, W. C. (1957). Somatic basis of sexual behavior patterns in guinea pigs: Factors involved in the determination of the character of the soma in the female. *Psychosom. Med.* 19: 114-151.

Goy, R. W., Bridson, W. E., and Young, W. C. (1964). Period of maximal suceptibility of the prenatal female guinea pig to masculinizing actions of testosterone propionate. *J. Comp. Physiol. Psychol.* 57: 166-174.

Goy, R. W., Phoenix, C. H., and Meidinger, R. (1967). Postnatal development of sensitivity to estrogen and androgen in male, female, and pseudohermaphroditic guinea pigs. *Anat. Rec.* 157: 87-96.

Green, R. (1974). *Sexual Identity Conflict in Children and Adults,* Basic Books, New York.

Green, R. (1975). Human sexuality: Research and treatment frontiers. In Brodie, H. K., and Hamburg, D. (eds.), *American Handbook of Psychiatry,* Vol. V6, Basic Books, New York.

Green, R., and Money, J. (eds.) (1969). *Transsexualism and Sex Reassignment,* Johns Hopkins University Press, Baltimore.

Green, R., and Stoller, R. (1971). Two monozygotic (identical) twin pairs discordant for gender identity. *Arch. Sex. Behav.* 1: 321-327.

Greer, G. (1971). *The Female Eunuch,* McGraw-Hill, New York.

Grϕnseth, E. (1971/1972). The husband provider role and its dysfunctional consequences. *Sociol. Focus* 5 (Winter): 10-18.

Gross, N., Mason, W. S., and McEachern, A. W. (1958). *Explorations in Role Analysis,* Wiley, New York.

Guttman, L., and Walsh, J. J. (1971). Prostigmin assessment of fertility in spinal man. *Paraplegia* 9: 39-51.

Hacker, H. M. (1951). Women as a minority group. *Social Forces* 30 (October): 60-69.

Hall, C., and Van Der Castle, R. L. (1966). *The Content Analysis of Dreams,* Appleton-Century-Crofts, New York.

Halverson, C. F., and Waldrop, M. R. (1973). The relations of mechanically recorded activity level to varieties of preschool play behavior. *Child Develop.* 44: 678-681.

Hatterer, L. J. (1970). *Changing Homosexuality in the Male,* McGraw-Hill, New York.

Hendricks, S. E. (1972). Androgen modification of behavioral responsiveness to estrogen in the male rat. *Horm. Behav.* 3: 47-54.

Hochschild, A. R. (1973). A review of sex role research. *Am. J. Sociol.* 78(4): 249-267.

Holter, H. (1970). *Sex Roles and Social Structure,* Universitetsforlaget, Oslo.

Hunt, M. (1974). *Sexual Behavior in the 1970s,* Playboy Press, Chicago.

Israel, J., Gustavsson, N., Eliasson, R. M., and Lindberg, G. (1970). Sexuelle Verhaltensformen der schwedischen Grossstadtjugend. In Bergstrom-Walan, M.-B., *et al.* (eds.), *Modellfall Skandinavien?* Rowohlt, Reinbek.

Jochheim, K. A., and Wahle, H. (1970). A study on sexual function in 56 male patients with complete irreversible injuries of the spinal cord and cauda equina. *Paraplegia* 8: 166-169.

Kanter, R. (1972). *Commitment and Utopia: Communes and Utopias in Sociological Perspective,* Harvard University Press, Cambridge, Mass.

Kinsey, A., Pomeroy, W., and Martin, C. (1948). *Sexual Behavior in the Human Male,* Saunders, Philadelphia.

Kinsey, A. C., Pomeroy, W. B., Martin, C. E., and Gebhard, P. H. (1953). *Sexual Behavior in the Human Female,* Saunders, Philadelphia.

Kohlberg, L. (1966). A cognitive-developmental analysis of children's sex-role concepts and attitudes. In Maccoby, E. E. (ed.), *The Development of Sex Differences,* Stanford University Press, Stanford, Calif.

Kolodny, R., Master, W., Hendryx J., and Toro, G. (1974). Depression of plasma testosterone levels after chronic intensive marihuana use. *New Engl. J. Med.* 290: 872-874.

Kotari, D. R., Timor, G. W., Frohrib, D. A., and Bradley, W. E. (1972). An implantable fluid transfer system for treatment of impotence. *J. Biometrics* 5: 567-570.

Kruez, L., Rose, R., and Jennings, J. (1972). Suppression of plasma testosterone levels and psychological stress. *Arch. Gen. Psychiat.* 26: 479-482.

Krusen, F. H., Kottke, F. J., and Ellwood, P. M., Jr. (eds.) (1971). *Handbook of Physical Medicine and Rehabilitation*, 2nd ed., Saunders, Philadelphia, p. 2.

Lash, H. (1968). Silicone implant for impotence. *J. Urol.* 100: 709-710.

Lebovitz, P. (1972). Feminine behavior in boys — Aspects of its outcome. *Am. J. Psychiat.* 128: 1283-1289.

Levine, S., and Mullins, R., Jr. (1964). Estrogen administered neonatally affects adult sexual behavior in male and female rats. *Science* 144: 185-187.

Levitt, E. E., and Klassen, A. D., Jr. (1973). Public attitudes toward sexual behaviors: The latest investigation of the Institute for Sex Research. Paper presented at the American Orthopsychiatric Association meeting (preliminary findings from study).

Lev-Ran, A. (1974). Gender role differentiation in hermaphrodites. *Arch. Sex. Behav.* 3: 339-424.

Lewis, M. (1967). The meaning of a response or why researchers in infant behavior should be oriental metaphysicians. *Merrill-Palmer Quart.* 13(1): 7-18.

Lewis, M. (1972a). Introduction to a symposium, cross-cultural studies of mother-infant interaction: Description and consequence. *Hum. Develop.* 15: 75-76.

Lewis, M. (1972b). Parents and children: Sex-role development. *School Rev.* 80(2): 229-240.

Lewis, M. (1972c). State as an infant-environment interaction: An analysis of mother-infant interaction as a function of sex. *Merrill-Palmer Quart.* 18: 95-121.

Lewis, M., and Als, H. R. (1975). The contribution of the infant to the interaction with his mother. Paper presented at the Society for Research in Child Development meetings, Denver, April 1975.

Lewis, M., and Brooks, J. (in press). Infants' social perception: A constructivist view. In Cohen, L., and Salapatek, P. (eds.), *Infant Perception*, Academic Press, New York.

Lewis, M., and Freedle, R. (1973). Mother-infant dyad: The cradle of meaning. In Pliner, P., Krames, L., and Alloway, T. (eds.), *Communication and Affect: Language and Thought*, Academic Press, New York, pp. 127-155.

Lewis, M., and Lee-Painter, S. (1974). An interactional approach to the mother-infant dyad. In Lewis, M., and Rosenblum, L. (eds.), *The Effect of the Infant on Its Caregiver: The Origins of Behavior*, Vol. I, Wiley, New York, pp. 21-48.

Lewis, M., and Rosenblum, L. (eds.) (1974). *The Effect of the Infant on Its Caregiver: The Origins of Behavior*, Vol. I, Wiley, New York.

Lewis, M., and Weinraub, M. (1974). Sex of parent X sex of child: Socioemotional development. In Friedman, R. C., Richart, R. M., and Van de Wiele, R. L. (eds.), *Sex Differences in Behavior*, Wiley, New York, pp. 165-189.

Lewis, M., and Wilson, C. D. (1972). Infant development in lower class American families. *Hum. Develop.* 15(2): 112-127.

Lewis, M., McGurk, H., Scott, E., and Groch, A. (1973). Infants' attentional distribution across two modalities. Unpublished manuscript.

Lindner, H. (1953). Perceptual sensitization to sexual phenomena in the chronic physically handicapped. *J. Clin. Psychol.* 9: 67-68.

Lipman-Blumen, J. (1973). Role de-differentiation as a system response to crisis: Occupational and political roles of women. *Sociol. Inquiry* 43(2): 105-129.

Lipman-Blumen, J. (1975). Vicarious achievement roles for women: A serious challenge for guidance counselors. *Personnel Guid. J.* 53: 680.

Lipman-Blumen, J. (in progress). A homosocial view of sex roles.

Lynn, D. (1969). *Parental and Sex Role Identification: A Theoretical Formulation*, McCutchan, Berkeley, Calif.

Maccoby, E. E. (ed.) (1966). *The Development of Sex Differences*, Stanford University Press, Stanford, Calif.

Maccoby, E. E., and Jacklin, C. N. (1974a). *The Psychology of Sex Differences*, Stanford University Press, Stanford, Calif.

Maccoby, E. E., and Jacklin, C. N. (1974b). Sex differences revisited: Myth and reality. Presented at the annual meeting of the American Educational Research Association, Chicago.

Masters, W. H., and Johnson, V. E. (1969). Undue distinction of sex. *New Engl. J. Med.* 281: 1422-1423.

Mead, M. (1935). *Sex and Temperament in Three Primitive Societies,* Mentor Books, New American Library, New York.

Mednick, M. T. S. (in press). Social change and sex role inertia: The case of the kibbutz. In Mednick, M. T. S., Tangri, S. S., and Hoffman, L. W. (eds.), *Women and Achievement: Motivational and Social Analyses,* Hemisphere Press, Washington, D.C.

Merton, R. K. (1949). *Social Theory and Social Structure,* Free Press, Glencoe, Ill.

Merton, R. K. (1957). The role-set: Problems in sociological theory. *Brit. J. Sociol.* 8(2): 106-120.

Messer, S. B., and Lewis, M. (1972). Social class and sex differences in the attachment and play behavior of the one-year-old infant. *Merrill-Palmer Quart.* 18(4): 295-306.

Michalson, L., and Brooks, J. (1975). Social labeling: The acquisition of semantic features. Paper presented at the Eastern Psychological Association meetings, New York.

Michalson, L., Brooks, J., and Lewis, M. (1974). Peers, parents, people: Social relationships in infancy. Unpublished manuscript.

Millet, K. (1970). *Sexual Politics,* Doubleday, Garden City, N. Y.

Mischel, W. (1966). A social-learning view of sex differences in behavior. In Maccoby, E. E. (ed.), *The Development of Sex Differences,* Stanford University Press, Stanford, Calif., pp. 56-81.

Mitchell, G. D., and Brandt, E. M. (1970). Behavioral differences related to experience of mother and sex of infant in the rhesus monkey. *Develop. Psychol.* 3(1): 149.

Money, J. (1967). Sexual problems of the chronically ill. In Wahl, C. W. (ed.), *Sexual Problems: Diagnosis and Treatment in Medical Practice,* New York Free Press, New York, pp. 226-287.

Money, J., and Ehrhardt, A. (1972). *Man and Woman, Boy and Girl,* Johns Hopkins University Press, Baltimore.

Money, J., and Pollitt, E. (1964). Cytogenetic and psychosexual ambiguity: Klinefelter's syndrome and transvestism compared. *Arch. Gen. Psychiat.* 11: 589-595.

Money, J., Hampson, J., and Hampson, J. (1955). An examination of some basic sexual concepts: The evidence of human hermaphroditism. *Bull. Johns Hopkins Hosp.* 97: 301-319.

Moss, H. A. (1967). Sex, age, and state as determinants of mother-infant interaction. *Merrill-Palmer Quart.* 13(1): 19-36.

Myrdal, G. (1944). Appendix 5, a parallel to the Negro problem. In *An American Dilemma,* Harper, New York.

National Institute of Mental Health Task Force on Homosexuality (1972). *Final Report and Background Papers,* DHEW Publication No. (HSM) 72-9116, U.S. Government Printing Office, Washington, D.C.

Neubeck, G. (1972). The myriad motives for sex. *Sex. Behav.* 2(7): 51-56.

Noble, R. G. (1973). The effects of castration at different intervals after birth on the copulatory behavior of male hamsters. *Horm. Behav.* 4: 45-52.

Oppenheimer, V. K. (1968). The sex-labelling of jobs. *Indust. Relations* 7(3): 219-34.

Parsons, T. (1942). Age and sex in the social structure of the United States. *Am. Sociol. Rev.* 7(5): 604-616.

Parsons, T., and Bales, R. F. (1955). *Family Socialization and Interaction Process,* Free Press, Glencoe, Ill.

Paup, Coniglio, L. P., and Clemens, L. G. (1972). Masculinization of the female golden hamster by neonatal treatment with androgen-estrogen. *Horm. Behav.* 3: 123-132.

Pfaff, D. (1970). Nature of sex hormone effects on rat sex behavior: Specificity of effects and individual patterns of response. *J. Comp. Physiol. Psychol.* 73: 349-358.

Pfaff, D., and Zigmond, R. E. (1971). Neonatal androgen effects on sexual and non-sexual behavior of adult rats tested under various hormone regimes. *Neuroendocrinology* 7: 129-145.

Phoenix, C. H. (1973). Ejaculation by male rhesus as a function of the female partner. *Horm. Behav.* 4: 365-370.

Phoenix, C. H., Goy, R. W., Gerall, A. A., and Young, W. C. (1959). Organizing action of prenatally administered testosterone propionate on the tissues mediating mating behavior in the female guinea pig. *Endocrinology* 65: 369-382.

Phoenix, C. H., Slob, A., and Goy, R. W. (1973). Effects of castration and replacement therapy on the sexual behavior of adult male rhesus. *J. Comp. Physiol. Psychol.*

Pirke, K. M., Kockott, G., and Dittmar, F. (1974). Psychosexual stimulation and plasma testosterone in man. *Arch. Sex. Behav.* 3: 577-584.

Prince, F., and Bentler, P. (1972). Survey of 504 cases of transvestism. *Psychol. Rep.* 31: 903-917.

Rebelsky, F., and Hanks, C. (1971). Fathers' verbal interaction with infants in the first three months of life. *Child Develop.* 42(1): 63-68.

Reiss, I. L. (1968). How and why American's sex standards are changing, *Trans-action* 5: 26-32.

Richardson, S. A. (1972). People with cerebral palsy talk for themselves. *Develop. Med. Child Neurol.* 14: 524-535.

Rose, R., Bowne, P., and Poe, R. (1969). Androgen response to stress. *Psychosom. Med.* 31: 418-436.

Rubinstein, E. A., and Parloff, M. (eds.) (1959). *Research in Psychotherapy*, American Psychological Association, Washington, D.C.,

Saghir, M. T., and Robins, E. (1973). *Male and Female Homosexuality: A Comprehensive Investigation*, Williams and Wilkins, Baltimore.

Schmidt, G., and Sigusch, V. (1970). Sex differences in responses to psychosexual stimulation by films and slides. *J. Sex Res.* 6: 268-283.

Schmidt, G., and Sigusch, V. (1971). *Arbeiter Sexualität*, Luchterhand, Berlin.

Schmidt, G., and Sigusch, V. (1972). Changes in sexual behavior among young males and females between 1960-1970. *Arch. Sex. Behav.* 2: 27-45.

Schmidt, G., and Sigusch, V. (1973). Women's sexual arousal. In Zubin, J., and Money, J. (eds.), *Contemporary Sexual Behavior*, Johns Hopkins University Press, Baltimore, pp. 117-143.

Schmidt, G., Sigusch, V., and Schäfer, S. (1973). Responses to reading erotic stories: Male-female differences. *Arch. Sex. Behav.* 2: 181-199.

Sears, R., Rau, L., and Alpert, R. (1965). *Identification and Child Rearing*, Stanford University Press, Stanford, Calif.

Sigusch, V., and Schmidt, G. (1973). *Jugendsexualität*, Enke, Stuttgart.

Sigusch, V., Schmidt, G., Reinfeld, A., and Wiedemann-Sutor, I. (1970). Psychosexual stimulation: Sex differences. *J. Sex Res.* 6: 10-24.

Simon, W., and Gagnon, J. H. (1967). The pedagogy of sex. *Saturday Rev.* 91: 74-76, 91.

Sorensen, R. C. (1973). *Adolescent Sexuality in Contemporary America*, World Publishing, New York.

Spiro, M. E. (1956). *Kibbutz, Venture in Utopia*, Schocken Books, New York.

Stern, J. J. (1969). Neonatal castration, androstenedione, and the mating behavior of the male rat. *J. Comp. Physiol. Psychol.* 69: 608-612.

Stewart, J., Pottier, J., and Kaczender-Henrik, E. (1971). Male copulatory behavior in the female rat after perinatal treatment with an antiandrogenic steroid. *Horm. Behav.* 2: 247-254.

Stoller, R. J. (1968). *Sex and Gender: On the Development of Masculinity and Femininity*, Science House, New York.

Stoller, R. J. (1973). *Contemporary Sexual Behavior*, Johns Hopkins University Press, Baltimore.

Swanson, H. H., and Crossley, D. A. (1971). Sexual behavior in the golden hamster and its modification by neonatal administration of testosterone propionate. In Hamburgh, M., and Barrington, E. J. W. (eds.), *Hormones in Development*, Appleton Press, New York, pp. 677-687.

Talmon, Y. (1965). Sex-role differentiation in an equalitarian society. In Lasswell, T. E., *et al.* (eds.), *Life in Society*, Scott, Foresman, Chicago, pp. 144-155.

Tiefer, L., and Johnson, W. A. (1971). Female sexual behaviour in male golden hamsters. *J. Endocrinol.* 51: 615-620.

Tiger, L. (1970). *Men in Groups*, Random House, New York.

Tresemer, D., and Pleck, J. (1972). Maintaining and changing sex-role boundaries in men (and women). Presented at conference on women: Resources for a Changing World, Radcliffe Institute, Cambridge, Mass.

Tsuji, I., Nakajima, F., Morimoto, J., and Nounaka, Y. (1961). The sexual function in patients with spinal cord injury. *Urol. Int.* 12: 270-280.

Turner, R. H. (1970). *Family Interaction*, Wiley, New York.

United States Department of Labor (1971). *1969 Handbook of Women Workers,* Women's Bureau, U.S. Government Printing Office, Washington, D.C.

United States President, 1969 (1973). Nixon, Richard M. Economic Report of the President. Transmitted to Congress, January 1973, together with the annual report of the Council of Economic Advisors, U.S. Government Printing Office, Washington, D.C., 301 pp.

Urbain, E. S., Maccoby, E. E., and Wright, G. (1974). A study of categories of mother-child and child-child interaction in same-sex and cross-sex pairs of 18-month-old children. Unpublished manuscript, Stanford University, Stanford, Calif.

Ward, I. L., and Renz, F. J. (1972). Consequences of perinatal hormone manipulations on the adult sexual behavior of female rats. *J. Comp. Physiol. Psychol.* 78: 349-355.

Weinberg, M. S., and Bell, A. P. (1972). *Homosexuality: An Annotated Bibliography,* Harper and Row, New York.

Weiss, A. J., and Diamond, M. D. (1966). Sexual adjustment, identification, and attitudes of patients with myelopathy. *Arch. Phys. Med. Rehab.* 47: 245-250.

Weiss, R. S. (1973). *Loneliness: The Experience of Emotional and Social Isolation,* M.I.T. Press, Cambridge, Mass.

Whalen, R. E. (1971). The ontogeny of sexuality. In Moltz, H. (ed.), *Ontogeny of Vertebrate Behavior,* Academic Press, New York, pp. 229-261.

Whalen, R. and Edwards, D. (1967). Hormonal determinants of the development of masculine and feminine behavior in male and female rats. *Anat. Rec.* 157: 173-180.

Whalen, R. E., and Luttge, W. G. (1971). Perinatal administration of dihydrotestosterone to female rats and the development of reproductive function. *Endocrinology* 89: 1320-1322.

Wickler, W. (1973). *The Sexual Code.* Anchor Press/Doubleday, Garden City, N.Y.

Yalom, I., Green, R., and Fisk, N. (1973). Prenatal exposure to female hormones — Effect on psychosexual development in boys. *Arch. Gen. Psychiat.* 28: 554-561.

Young, W., Goy, R., and Phoenix, C. (1964). Hormones and sexual behavior. *Science* 143: 212-218.

Young, W. C. (1961). Hormones and mating behavior. In Young, W. C. (ed.), *Sex and Internal Secretions,* Williams and Wilkins, Baltimore, pp. 1173-1239.

Young, W. C. (1969). Psychobiology of sexual behavior in the guinea pig. *Advan. Stud. Behav.* 2: 1-110.

Young, W. C., and Rundlett, B. (1939). The hormonal induction of homosexual behavior in the spayed female guinea pig. *Psychosom. Med.* 1: 449-460.

Zeitlin, A. B., Cottrell, T. L., and Lloyd, F. A. (1957). Sexology of the paraplegic male. *J. Fertil. Steril.* 8: 337-344.

Zetterberg, H. L. (1969). *Om Sexuallivet i Sverige,* Statens offentliga utredninger, Stockholm.

Zuger, B. (1966). Effeminate behavior present in boys from early childhood. *J. Pediat.* 69: 1098-1107.

Zuger, B. (1970). Gender role differentiation: A critical review of the evidence from hermaphroditism. *Psychosom. Med.* 32: 449-463.

Conference Participants

Alan P. Bell, Ph.D.
Institute for Sex Research
Indiana University
Bloomington, Indiana

Theodore M. Cole, M.D.
Department of Medicine
University of Minnesota
Minneapolis, Minnesota

Gerald Davison, Ph.D.
Department of Psychology
State University of New York
  at Stony Brook
Stony Brook, New York

Diane S. Fordney-Settlage, M.D.
Department of Obstetrics and
  Gynecology
School of Medicine
University of Southern California
Los Angeles, California

Paul H. Gebhard, Ph.D.
Institute for Sex Research
Indiana University
Bloomington, Indiana

James Geer, Ph.D.
Department of Psychology
State University of New York
  at Stony Brook
Stony Brook, New York

Robert Geiger, M.D.
Department of Orthopedic Surgery
University of California Medical Center
San Francisco, California

Norman Goodman, Ph.D.
Department of Sociology
State University of New York
  at Stony Brook
Stony Brook, New York

Robert W. Goy, Ph.D.
Department of Psychology
University of Wisconsin
Madison, Wisconsin

Richard Green, M.D.
Department of Psychiatry and
  Behavioral Science
State University of New York
  at Stony Brook
Stony Brook, New York

Jennifer James, Ph.D.
Department of Psychiatry
University of Washington
Seattle, Washington

Michael Lewis, Ph.D.
Educational Testing Service
Princeton, New Jersey

Jean Lipman-Blumen, Ph.D.
Office of Research
National Institute of Education
Washington, D.C.

Joseph LoPiccolo, Ph.D.
Department of Psychiatry and
  Behavioral Science
State University of New York
  at Stony Brook
Stony Brook, New York

John Messenger, Ph.D.
Department of Anthropology
Ohio State University
Columbus, Ohio

Robert M. Rose, M.D.
Department of Psychosomatic Medicine
Boston University School of Medicine
Boston, Massachusetts

Eli A. Rubinstein, Ph.D.
Department of Psychiatry and
  Behavioral Science
State University of New York
  at Stony Brook
Stony Brook, New York

Gunter Schmidt, Ph.D.
Institute for Sex Research
University of Hamburg
Hamburg, West Germany

John Vandenburgh, Ph.D.
North Carolina Department of
  Mental Health
Research Division
Raleigh, North Carolina

Stanley F. Yolles, M.D.
Department of Psychiatry and
  Behavioral Science
State University of New York
  at Stony Brook
Stony Brook, New York

Official Observers:

Jack Wiener, NIMH
George Renaud, NIMH
Edward M. Brecher, writer